D1336201

About

This book is a guide to the wonders of the sky and also to the serious student. Using it, even the complete beginner ... there are maps of the Moon, lists of interesting objects, and—for the classically-minded — mythological notes about constellation names. Even with the naked eye there is much to see in the night sky, and there are also possibilities of carrying out interesting research—for instance, in observing meteor showers. The *New Observer's Book of Astronomy* deals with this in full, and also with the needs of the owner of binoculars or an astronomical telescope.

The book is not designed solely for readers in the northern hemisphere, and is of equal use to those living south of the equator.

The first edition of this book appeared in 1962, and has since been reprinted six times, with revisions. For this new edition (the seventh) the text has been completely updated and much new material has been added.

About the Author

Patrick Moore's interest in astronomy dates from the age of six. When still a schoolboy he joined the British Astronomical Association, and is its current President. During World War II he served as a navigator with RAF Bomber Command. He subsequently set up his private observatory, first in East Grinstead and then, after a spell of three years as Director of the Armagh Planetarium (1965–8) in his present home at Selsey.

He has made many contributions to astronomy, mainly in connection with the Moon and planets, and is the author of many technical papers as well as popular books and articles. In April 1957 he began his monthly series on BBC Television, *The Sky at Night*, which has continued uninterrupted ever since—a world television record. He has received the Jackson-Gwilt Medal of the Royal Astronomical Society, the Goodacre Medal of the British Astronomical Association, the Roberts-Klumpke Medal of the Astronomical Society of the Pacific and various other awards; he is an Honorary Member of the Astronomical Society of the USSR. A minor planet, Asteroid 2602, has been named 'Moore' in his honour. He was awarded the OBE in 1967 for his services to astronomy.

The New Observer's Book of

Astronomy

Patrick Moore

O.B.E., D.Sc. (Hon.), F.R.A.S.
President of the British Astronomical Association

With 32 plates
In colour and black and white
Including 10 specially drawn
by L. F. Ball

Frederick Warne

Published by Frederick Warne (Publishers) Ltd.
London, England

© Sixth Revised Edition 1978
Reprinted 1979, 1983 (revised format)
Originally published as *The Observer's Book of
Astronomy* in small hardback format

Library of Congress Catalog Card No. 62–18908

ISBN 0 7232 1575 8

Printed and bound in Great Britain
by William Clowes (Beccles) Limited,
Beccles and London

CONTENTS

LIST OF PLATES

ACKNOWLEDGMENTS

Acknowledgments are due to the Science Museum, London, for permission to reproduce the following illustrations:

Plate 5 (lower half). Crown copyright, Science Museum, London.

Plates 29 and 8: Photo. Science Museum, London.

Plate 28: Photo. Science Museum, London. By courtesy of Mount Wilson Observatory, California.

Plates 24 and 25: Photo. Science Museum, London. By courtesy of l'Observatoire de Paris.

Plate 1: Lent to the Science Museum, London, by the late J. Franklin Adams, Esq., F.R.A.S.

Acknowledgments are also due to the following sources: *Plate 26* is a Lick Observatory photograph, and the photograph reproduced in *Plate 16* was obtained at the Lowell Observatory by E. C. Slipher.

Plate 22 is taken from a photograph by Mr W. M. Baxter. I am indebted to W. H. Barnes for *Plate 31*.

Plates 19, 20 and 32 are reproduced by permission of NASA.

My grateful thanks are also due to L. F. Ball, who has drawn plates especially for this volume.

INTRODUCTION

During the last decade there has been a tremendous surge of popular interest in astronomy. Undoubtedly the space-rockets have had something to do with this; but the interest has always been there, and for good reasons.

Astronomy is one of the few sciences in which the amateur may still make himself useful. His field of research is naturally limited, but is wider than might be thought. For instance, professional astronomers pay relatively little attention to the surface features of the Moon and planets. The world's largest telescopes are engaged in more fundamental studies, and so the amateur who decides to turn himself into a serious observer is always warmly welcomed. The main requirements are enthusiasm and patience.

This book, however, is written not for the serious observer, but for the beginner—the man who knows little or nothing about astronomy, but who is interested enough to want to learn the main facts. Some equipment is desirable, but the cost need not be great, and many of the features here may be well seen with modest binoculars.

Amateur astronomers are of all ages, and come from all walks of life. T. E. R. Phillips, world-famous for his studies of the planet Jupiter, was a country clergyman, while H. P. Wilkins worked by day in the Ministry of Supply in London, and spent his spare time in compiling a large and detailed map of the Moon. W. T. Hay, who discovered the white spot on Saturn in 1933, was better known to the general public as Will Hay the stage and film comedian; W. F. Denning, who was awarded the Gold Medal of the Royal Astronomical Society for his planetary and meteor work, was an accountant by profession. These instances are typical of a great many others.

The first step to be taken by the would-be observer is to acquaint himself with the fundamental principles of astronomy, which may be done by reading books. Next, probably, comes star recognition, which is not nearly so difficult as it may sound; once the main constellation patterns have been identified, the remainder fall into place without a great deal of trouble. By the time he can find his way around the sky, our beginner will have decided which branch of observation interests him most, and he will be ready to turn to some more technical reference book.

The professional astronomer is necessarily a mathematician. Obviously mathematical knowledge will be of tremendous help to the amateur also, but in some fields of research it is not essential, and the non-mathematician is not debarred from playing an active part in practical observation.

So let us begin at the very beginning, and outline the basic facts of astronomical science.

THE STARLIT SKY

Early men believed the Earth to be flat, and to lie at rest in the centre of the universe, with the Sun and all other celestial bodies moving round it once a day. Such beliefs were natural enough, but long before the time of Christ it became known that the Earth is not a flat plane; it is a globe, almost 8,000 miles in diameter. During the great period of classical Greece, Eratosthenes of Cyrene measured the Earth's circumference with surprising accuracy, and the value which he gave was actually better than that used by Christopher Columbus nearly 2,000 years later.

Aristarchus of Samos, who lived between 310 and 230 B.C., believed that the Earth must revolve round the Sun, completing one journey per year. His ideas were not accepted by his contemporaries, and later Greek scientists returned to the idea of an Earth-centred universe, an error which held up scientific progress for hundreds of years. The true state of affairs was not demonstrated until the sixteenth and seventeenth centuries, when the work of men such as Copernicus, Kepler and Galileo made it clear that the Earth is, after all, nothing more than an ordinary planet.

The *Solar System* in which we live consists of one star (the Sun), nine planets and various bodies of lesser importance, such as satellites and comets. Of the planets known to the ancients, Mercury and Venus are closer to the Sun than we are, while Mars, Jupiter and Saturn are further away. Three still more distant planets have been discovered,

Uranus (1781), Neptune (1846) and Pluto (1930). Uranus may just be seen with the naked eye under suitable conditions, but Neptune and Pluto are invisible without optical aid.

The planets have no light of their own, and shine only because they reflect the rays of the Sun. Their revolution periods range from 88 days for Mercury to 248 terrestrial years in the case of Pluto. They are comparatively close to us—Venus at its nearest is less than 25,000,000 miles away—and so appear much more imposing than they really are. Mars, for instance, may shine with a brilliance far exceeding that of Sirius, the brightest star. Yet Mars is a small world with a diameter only half that of the Earth, while Sirius is a sun in its own right, 26 times as luminous as our Sun.

The nearest of the celestial bodies is, of course, the Moon, which lies at a distance of less than 250,000 miles, and is the Earth's only natural satellite. To us it appears as the most splendid object in the sky apart from the Sun; in reality it is a very junior member of the Solar System. Some of the largest planets have veritable satellite families. Jupiter is attended by no less than 16 moons, while Saturn has at least 20.

The distance between the Earth and the Sun, 93,000,000 miles, is known as the *astronomical unit*. Beyond the Solar System, however, we find that both the mile and the astronomical unit are inconveniently small for distance reckoning; the nearest of the 'night' stars is some 25,000,000,000,000 miles away, and most of the others are much more remote. Consequently a new unit is used, the *light-year*. Light travels at 186,000 miles per second; in a year, therefore, it covers, 5,880,000,000,000 (nearly six million million) miles, and this is the astronomer's 'yard-stick'. The nearest star lies at a distance of rather more than 4 light-years.

The fact that the stars are so remote means that their individual or *proper motions* are very slight, even though they are really moving through space at very high speeds. A man of Julius Cæsar's time must have seen the star-groups in virtually the same form as now; even to the builders of the Egyptian pyramids there can have been no obvious differences. The ancients too will have noticed the seven stars which make up the characteristic Plough, as well as the brilliant constellation Orion and all the other groups with which we are familiar. In fact the constellations appear unchanging for year after year, century after century, and we can well understand why our ancestors spoke about the so-called 'fixed stars'.

Over a sufficiently long period, the stellar proper motions will accumulate sufficiently to alter the patterns. By the year A.D. 200,000, for instance, the Plough will have lost its present outline, since two of its stars (Alkaid and Dubhe) are moving through space in a direction different from that of the remaining five members. Modern instruments are capable of measuring the annual proper motions of most of the bright naked-eye stars. These shifts, however, are too tiny to concern the amateur observer, who may for most purposes regard the constellation forms as permanent.

The planets, so much closer than the stars, seem to wander around the sky; in fact the word planet means 'wandering star'. During 1967 Jupiter lay in the constellation of Cancer, the Crab; in 1968 it had moved into Leo, the Lion, and so on. All the main planets keep to a well defined belt across the sky, marked by the twelve constellations of the Zodiac. Careful checking of the positions of the planets from night to night will soon show that they are in motion.

The daily movement of the heavens from east to west is due to the rotation of the Earth on its axis, and has nothing to do with the stars themselves. In a northward direction the axis points toward the *north celestial pole*, which is marked approximately by the bright star Polaris. Polaris therefore seems to remain almost stationary, while all the other bodies (including the Sun, Moon and planets) appear to revolve round it once in 24 hours. From southern latitudes Polaris cannot be seen, since it never rises above the horizon; there is no conspicuous south polar star, the nearest naked-eye object being a faint star known as Sigma Octantis.

The apparent positions of celestial bodies are measured in terms of *declination* and *right ascension*. Declination, which corresponds to 'sky latitude', is expressed as angular distance north or south of the celestial equator. It is convenient to imagine that the sky is solid, so that lines may be drawn on it; if we project the plane of the Earth's equator on to this *celestial sphere*, we naturally have the celestial equator. Stars on the equator have declination $0°$; the north celestial pole has, of course, declination $+90°$, or $90°$ north. Polaris, with a declination of $+89° 2'$, is therefore less than one degree from the polar point.

The altitude of the celestial pole is always equal to the latitude of the observer. London, for example, has a latitude of roughly $51\frac{1}{2}°$ N.; the pole will therefore be $51\frac{1}{2}°$ above the horizon. For the sake of simplicity let us take the latitude as being $52°$ and see what information may be drawn from it.

If we subtract 52 from 90, we obtain 38. This means that any object with a declination north of

+ 38° will never set; it will remain above the horizon all the time, and will be circumpolar. On the other hand, an object with a declination south of − 38° will never rise at all, and will never be seen. Thus the bright southern star Canopus, with a declination of − 53°, is permanently invisible from Britain. If we want to observe it we must travel to more southerly lands.

The celestial equator acts as our reference line for measurement of declination, but for east–west reckoning we need another standard also. This is provided by the *ecliptic*, which is defined by the plane of the Earth's path or *orbit*. The ecliptic intersects the celestial sphere in a great circle, and this will also be the apparent path of the Sun among the stars in its yearly journey around the sky. Since the plane of the Earth's equator is tilted at an angle of about 23½° to that of the orbit, the ecliptic lies at an angle to the celestial equator. About 21st March each year the Sun reaches the celestial equator, travelling from south to north; this is where the ecliptic and the equator cross, and is called the First Point of Aries or *Vernal Equinox*. Here is our 'prime meridian' of the sky, and star positions measured from it in an eastward direction along the celestial equator are measures of the star's right ascension. Clearly, then, right ascension corresponds to longitude on the Earth's surface. It is convenient to measure eastward, though it would be equally possible to follow the system adopted for measuring terrestrial longitude.

Unlike declination, right ascension is generally reckoned in units of time. A star is said to culminate when it reaches its highest point above the observer's horizon, and is in his meridian; the First Point of Aries, which is not marked by any conspicuous star, therefore culminates once a day

15

(often, of course, when the Sun is above the horizon). The right ascension is the time-difference between the culmination of the First Point of Aries and the culmination of the star concerned. Rigel, the brilliant white star in Orion, culminates 5 hours 12 minutes after the First Point has done so; therefore the right ascension of Rigel is 5 hours 12 minutes.

Owing to an effect known as precession, due to a small annual shift in position of the celestial pole, the right ascensions and declinations of stars change slightly from year to year. These differences are, however, so small that for most purposes they may be neglected. On the other hand the right ascensions and declinations of the Sun, Moon and planets change considerably even over a period of a few hours.

An interesting and very simple experiment, which may be carried out with an ordinary camera, is to make a time-exposure of the starlit sky. The result will be a series of trails due to the shifting of the stars across the heavens. If the camera is directed toward the polar regions of the sky, the effect will be very striking, as is shown in Plate 1.

The star-system or *Galaxy* is made up of approximately 100,000,000,000 suns, many of them far larger and more luminous than our own. Rigel in Orion is an example of a real 'celestial searchlight', since it shines with 60,000 times the candlepower of the Sun, and is over 900 light-years away,[1] so that we are now seeing it as it used to be about the time of the Battle of Hastings. On the other hand we also know of very small and feeble stars, which we may regard as the glow-worms of the Galaxy.

---

[1] This is the latest estimate: the previous value was 540 light-years.

The stars are arranged in a flattened system, with a pronounced central nucleus; it has been said that the shape resembles that of two fried eggs laid back to back! The whole Galaxy measures 100,000 light-years from side to side, and the Sun is situated about 32,000 light-years from the centre. When we look along the main plane of the system, we see many stars in more or less the same direction, and this produces the effect known as the Milky Way. During a dark night the Milky Way is a glorious spectacle, and appears as a luminous band stretching across the sky. It is composed of stars, and these seem to be packed closely together, though in reality the Milky Way stars are not in the least crowded.

All the bright naked-eye stars belong to our own particular Galaxy, but still we are only at the beginning of things. Far away in space we can make out other galaxies, which are separate systems equal in status to ours. Three of them are visible without optical aid. Two of these, the twin Nubeculæ or Clouds of Magellan, have high southerly declination and cannot be seen from Britain; the third, the Great Spiral in Andromeda, may be glimpsed as a faint hazy patch of light. Binoculars show it very clearly, though large instruments are required for it to be seen in its full glory. The world's most famous telescope, the 200-inch reflector at Palomar in California, is capable of photographing 1,000,000,000 galaxies. The most distant systems so far recorded are several thousand million light-years away.

The universe is indeed a large place, and there is endless scope for the amateur astronomer as well as for the professional. Moreover the heavens are constantly changing, and there is always something new to see.

EQUIPMENT

Some branches of astronomical observation, such as meteor studies and records of auroræ, are best carried out with the naked eye. Generally, however, the amateur will need a telescope of some sort, and this is often where difficulties begin. The question most frequently asked by the beginner is: 'Where can I buy a cheap astronomical telescope which will allow me to make a real start in observing?'

Unfortunately this is a question which cannot be answered. The plain fact is that no good, cheap telescope is on the market. A 3-inch refractor, which used to cost less than £10 before the war, is now priced at something nearer £250; reflectors of sufficient aperture are not, basically, so expensive, but are difficult to find. Now and then, of course, excellent instruments may be found second-hand, but this is largely a matter of luck.

Of the two types of atronomical telescope, the more familiar is the refractor, which works in just the same way as a hand-telescope used for looking at ships out to sea. The light from the Moon, or whatever body is to be studied, passes through a glass lens or *object-glass*, and is brought to focus; the resulting image is then magnified by a smaller lens or *eyepiece*. The distance between the object-glass and the focal point is known as the *focal-length*.

A 3-inch refractor—that is to say, an instrument with an object-glass 3 inches in diameter—is the ideal telescope for the beginner. It is adequate to show features such as the craters and mountains of the Moon, the rings of Saturn and the satellites of Jupiter; it brings many hundreds of coloured stars, double stars, clusters and nebulæ within range, and useful work may be undertaken with it. Moreover it is portable and easy to handle, since it will not require adjustment from one year to another. Smaller refractors are of limited value astronomically, while larger ones are generally too heavy to be carried about, so that they have to be set up in a permanent position—preferably housed in an observatory. Anything larger than a 4-inch refractor must be regarded as strictly non-portable.

Light, as Newton demonstrated, is compound; what we normally term 'white' light is really a blend of all the hues of the rainbow, from violet to red. If a beam of sunlight be passed through a prism, it will be split up into a *spectrum*, because the prism will bend or refract the different colours by different amounts. Red light, for instance, is refracted less than blue. Unfortunately, the object-glass of a refractor tends to act in the same way, so that the blue and the red are brought to focus at slightly different points. This means that a bright object such as a star will be seen together with irritating false colour, which has nothing to do with the star itself but is produced by the telescope.

A partial remedy is to make the object-glass a compound arrangement, composed of two or more lenses fitted together; these lenses are made of different types of glass (crown and flint glass, for instance) whose errors tend to cancel each other out. The false colour trouble can never be completely cured, but it can be very much reduced. The object-glass of a refractor is therefore not the simple affair which it might be thought, and to make a good object-glass is beyond the powers of the average amateur. If you want a proper astronomical refractor, the only solution is to buy it ready-made.

Home-made refractors may, however, be constructed out of spectacle lenses and cardboard tubes. The method is to buy a spectacle lens of about 2 inches in diameter and 2 or 3 feet focal length, and also a smaller lens of shorter focal length, to act as an eyepiece. The lenses are then fitted into tubes—the eyepiece into a tube which slides in and out of the main one—and the result will be a telescope which is at least usable. It will be very limited; it will give an alarming amount of false colour, and it will certainly be less effective than good binoculars. However, it will be decidedly better than nothing, and it can be made in an afternoon at a cost of only a pound or so. The only trouble lies in buying the lenses, which is less easy now than it used to be some years ago.

One essential point to remember is that even a home-made telescope must have a stand, preferably a tripod similar to that used by a photographer; otherwise the instrument will be virtually useless. It is bound to have a small field of view, and to hold it sufficiently steady is to all intents and purposes impossible.

The question of a mounting for a professionally made refractor is equally important. Three-inch telescopes are often sold with the notorious pillar-and-claw stand, which looks beautifully neat but is as unsteady as a jelly. A tripod is much to be preferred, and should be as massive as possible. For astronomical work it is essential to keep the telescope completely rigid. Luckily, almost all pillar-and-claw mounted refractor telescopes may be removed and put on to tripods without much difficulty.

The function of an object-glass is to collect light; the actual work of magnification is done by the eyepiece, and a 3-inch refractor will need several eyepieces. For star-fields and general views a low power (say 30 diameters, or × 30) is desirable; for more detailed views, something in the region of × 80; and for studying the Moon and planets under really good conditions it is as well to have a relatively high power of around × 120. If the object glass is really good it may be possible to use still greater magnifications, but there are dangers here. It is a cardinal rule of observation that the image must be sharp and clear-cut; if there is the slightest blurring or unsteadiness, it will be wise to change at once to a lower power. The increased sharpness will more than compensate for the loss of actual magnification.

Eyepieces are made to a standard thread, so that any eyepiece will fit any telescope—theoretically, at least. They are not particularly expensive, and once the amateur has equipped himself with a few suitable eyepieces he will have to spend no more money on that score.

One often reads advertisements of very small refractors ($1\frac{1}{2}$- or 2-inch object-glasses) made originally for terrestrial use. It is said that such telescopes are well suited for astronomical purposes, and that they will give excellent results. My only comment here is that I have yet to find such a telescope which is of the slightest use in astronomy. Rather than buy one the beginner would be far better advised to invest in a pair of binoculars.

Incidentally, an astronomical telescope gives an inverted image, with north at the bottom and south at the top. With an instrument made for terrestrial use an extra lens system is inserted to correct this and put the image the right way up again. Yet every time light passes through the lens a little of it is lost, and the one thing that the astronomer is anxious to avoid is loss of light. The extra lenses are therefore left out, and many telescopic pictures in astronomy, whether drawings or photographs, are oriented with south at the top.

In short, the amateur who wants to own a refractor has the choice of buying a 3-inch or larger instrument, at considerable cost, or else making a very small telescope out of spectacle lenses and cardboard tubes, which will naturally have a very limited range. Small refractors meant originally for terrestrial use are best avoided. Before turning to the problems of larger telescopes, let us look at the second type of instrument—the reflector. Here we have no object-glass at all; the light is collected by means of a specially shaped mirror.

In the Newtonian, which is the usual form (so called because the principle was first developed by Sir Isaac Newton around 1671), the light passes down an open tube until it falls upon the main mirror or *speculum*. This mirror is curved, and reflects the light back up the tube until it falls on to a smaller mirror or *flat*, placed at an angle of 45°. The light is then directed to the side of the tube, where an image is formed and is magnified by an eyepiece as before. With a Newtonian reflector, then, the observer looks into the tube instead of up it.

Older mirrors were made of metal; modern ones are made of glass—it is very doubtful whether many metal mirrors are still in use. In either case the upper surface is coated with a layer of highly reflective material, either silver or aluminium. Silvering is cheaper and may be done at home, but it does not last for so long, and on the whole it is probably better to have the speculum professionally aluminized. The cost is not high, and the process need not be carried out very often.

A reflector has several advantages over a refractor. For one thing a mirror reflects all parts of the spectrum equally, so that there is no false colour trouble (apart from that inherent in the eyepiece, which is minor). Moreover a mirror is cheaper than a lens of equal light-grasp, and it can be home-made. On the other hand a reflector is more trouble; the mirrors can easily go out of adjustment, and need much more careful attention.

Inch for inch, a mirror is less effective than a lens. As we have seen, a 3-inch refractor is an excellent instrument for the beginner who wants to start proper observation, but a 3-inch reflector would be of very limited use. The minimum useful size for a reflector is 4 inches, and an instrument with at least a 6-inch mirror is highly desirable. Here again, unhappily, the would-be-purchaser of a new 6-inch reflector is faced with considerable financial outlay; £250 is a conservative estimate, though cheaper second-hand instruments are to be found now and again. It must also be added that extreme caution is needed before making a purchase. A reflector may well look imposing when it has, in reality, a badly figured mirror which will give no proper results. Bad mirrors cannot be detected by a casual look, and the beginner who buys a reflector without having the optical system thoroughly tested by an expert is asking for trouble. (With the utmost regret, it must also be recorded that even new reflectors are not always above suspicion. I once tested a very neat, workmanlike-looking 4-inch marketed at about £20. It proved to be quite unusable except with a very low magnification.)

Mirror-grinding is an art, but is more laborious

23

than genuinely difficult. The procedure is to buy two glass 'blanks' of the required size (say 6 inches), one to form the mirror and the other to be used as a 'tool'. By grinding one against the other, and carrying out periodical optical tests with simple apparatus, it is possible to produce a mirror with a satisfactory optical curve. The method is beyond the scope of this book, but several excellent descriptions are available. I know of one 16-year-old enthusiast who has made himself an excellent 8-inch without any outside help.

The blanks are not expensive—a few pounds will suffice—and the flat is not a major item financially. On the other hand, it is only fair to say that the beginner who sets out to grind a mirror is bound to come across innumerable snags, and will probably have several failures before producing a workable mirror. Also the grinding takes a long time, and there are no short cuts.

The tube of a reflector is needed only to hold the optical components in the correct positions, and there is no reason to make it round; a square tube, made perhaps of wood, is just as effective. Skeleton tubes, too, are popular, and have some advantages. A 6-inch skeleton reflector—a telescope with an open framework tube—is portable enough, whereas a 6-inch refractor is bound to be a very massive affair indeed.

Before many years are past, some firm or other is bound to start producing good 6-inch reflectors at a price which is within the range of the average amateur. Until the arrival of that happy time the beginner is best advised to resign himself to spending several tens of pounds; to wait until he can find a good second-hand instrument, which may take him years; or else—and best of all—to make his own.

A reflector, too, needs a really rigid mounting, and this may be the moment to make a few remarks about telescope stands in general.

Owing to the Earth's rotation on its axis, the celestial bodies seem to move steadily across the sky. Their directions and altitudes change constantly, as everyone knows. The movement is too slow to be noticed with the unaided eye except over periods of several minutes; but when a telescope is used the movement is magnified. With a high power a celestial object will seem to move very rapidly, and the observer will have to keep shifting the telescope in order not to lose sight of the body which he is studying. The speed with which a star or planet races across a telescopic field never fails to intrigue the novice who sees it for the first time.

Obviously this is highly inconvenient when detailed studies are being made. To make matters worse, with an ordinary *altazimuth* stand, such as a simple tripod, the telescope has to be moved both east to west, to compensate for the changing direction, and up or down, to compensate for the changing altitude. With a 3-inch refractor it is practicable to move the instrument merely by hand adjustments, but even so it is not always easy to keep an object in view when the most powerful eyepieces are being used. *Slow motions* may be fitted to altazimuth mounts for both refractors and reflectors; in such a case the movements are controlled by rotating rods. The main trouble with an arrangement of this kind is that the observer needs three hands.

Much more convenient is the *equatorial* stand. This time the axis of the mount points to the celestial pole, so that only the direction has to be allowed for by moving the telescope; the changing altitude will look after itself. The enthusiast who decides to make his own mounting will be wise to choose an equatorial. It is not much more complicated than an altazimuth, and is far better in practice.

Manual slow motions are more or less essential for an astronomical telescope of aperture greater than 3 inches. The ideal, of course, is to fit a 'clock drive' which will move the telescope steadily so that it follows the object under view. The problem is purely a mechanical one, and is not likely to present real difficulty to the amateur who is good with his hands. The clock may be weight-driven, though electric drives are now more common.

Another easy refinement to help in observation is to fit the main telescope with a *finder*. This is a much smaller telescope, almost always a refractor, mounted on the main tube. To find an object such as a star or planet is not easy with a high magnification; the finder, with its small aperture, will have a large field of view, and can at once be turned toward its target. The star is then brought into the middle of the finder-field, and if the instruments are correctly lined up the star will be found to be visible in the eyepiece of the main telescope. It is sometimes said that finders are unnecessary. This may be true; but they are cheap and easy to fit, and are certainly convenient. For a 3-inch refractor or a 6-inch reflector even a toy telescope suffices, though the serious amateur will probably want something with a greater light-grasp.

So far we have been dealing mainly with small telescopes which are easy to move around, and may be kept indoors or in a garden shed. This is often convenient, since it is generally found that any neighbouring trees and houses lie in the most awkward positions possible, and seldom fail to hide the particular object which the observer is anxious to study. (It is not so easy to move an equatorial stand *en bloc*, since before use the axis has to be lined up with the celestial pole, and in this respect it must be admitted that the altazimuth has its advantages.) With larger telescopes new problems naturally arise.

Lenses and mirrors are sensitive things, and must always be treated with great respect. They must be kept as dust- and moisture-free as possible, so that when not in use they have to be covered. It is easy to make a cap for the object-glass of a refractor; with a 3-inch, a cocoa-tin lid lined with blotting-paper will suffice. With a reflector, it is desirable to make effective covers for both mirror and flat. Before observing, remember to uncap the flat before exposing the main mirror, otherwise there is always a chance that the flat cover will fall down the tube and hit the glass, doing an incredible amount of damage in less than a second. The better your covers, of course, the less often will your mirrors have to be resilvered or aluminized.

When a telescope has to be set up in a permanent position, the first essential is to make sure that the maximum sky view is obtained, and to take account of all local hazards such as trees. Since rigidity is so important, it is wise to set the main pillar in concrete. Next comes the problem of protecting the telescope from wind, rain and damp.

For a refractor, or for a solid-tube reflector, a car cover will serve. It is far from ideal, and constant checks must be made on the moving parts of the mount; also it is essential to make sure that the whole optical system is well capped. Improvised shelters may also be built, but it is clearly better to make an observatory of some sort.

The conventional observatory takes the form of a domed building with a rotating roof. A wide slit in the roof is detachable, or may be folded back; when the telescope is to be used, the dome is simply moved round until the slit is in a suitable position. Although the observatory cannot be heated, because this would cause atmospheric turbulence, it makes conditions very comfortable. The observer is protected from the wind, and so is the telescope, so that no 'shake' can be produced from this cause.

Building an observatory is a fairly major operation, and the amateur will often find that no suitable site is available. It should be added, too, that the much popularized 'roof-top observatory' is not to be recommended. During winter, at least, heated air will rise from the house below, and make proper observing impossible owing to atmospheric turbulence. It is far better to build the observatory at ground level, as far from the dwelling-house as is practicable.

If a dome is out of the question, it may nevertheless be possible to construct a run-off shed. My own 12½-inch reflector is housed in a shed which is in two parts, and is mounted on rails. Before observing, the shed is unlatched, and the two parts simply pushed back in opposite directions, leaving the telescope exposed. A skilful carpenter can make a building of this sort cheaply, since the only cost will be that of the wood (or perhaps hardboard), the rails and eight small wheels. Alternatively, of course, the shed can be a single structure with a door at one end, so that it may be pushed clear in one piece. I used to have a shed of this sort, but found it rather cumbersome and heavy to move.

Other designs may be produced; it is possible, for instance, to have a low wall and a relatively lightly constructed roof which lifts off, or a roof which slides right back. Everyone will have his own ideas on the subject. I personally favour the two-piece shed on rails, having used it for many years and found it eminently satisfactory.

Fully equipped observatories are of course prohibitively expensive, but it is always worth while paying a visit to one. Unfortunately there are few public observatories in Britain (though America is better served), and it is a pity that opportunities are so restricted.

The most famous telescope in the world is the 200-inch reflector at Palomar, in California. The immense light-grasp of this instrument has led to tremendous advances in our knowledge of the stars and star-systems, but—like other great telescopes— it is not often used for studies of the Moon or planets. This policy is entirely justifiable, since lunar and planetary work may be carried through with smaller instruments. (Of course, results from space-probes are now all-important in Solar System studies.)

Other large reflectors include the 120-inch at Lick Observatory, also in America, and the 100-inch at Mount Wilson, which was erected in 1917 and remained unsurpassed until the completion of its neighbour at Palomar. A 98-inch reflector was set up at Herstmonceux in Sussex, the new site of the famous Royal Greenwich Observatory, but it has now been moved to a better site in La Palma (Canary Islands); and a new 236-inch reflector has been erected in the Soviet Union.

In recent years there has been an emphasis on southern-hemisphere telescopes, partly because of the excellent conditions in some southern countries and partly because there are some particularly important objects (such as the Magellanic Clouds)

29

which are inaccessible from Europe or the United States. Reflectors of between 150 and 160 inches aperture have been set up at Cerro Tololo (Chile), La Cilla (also Chile), Siding Spring (Australia) and Mount Stromlo (Canberra, also Australia). It must be said that these telescopes are more 'modern' than the Palomar 200-inch, and the same is true of the 158-inch reflector which has been set up at Kitt Peak in Arizona.

Large lenses are more difficult to make than large mirrors of equal light-grasp. Moreover a lens must be supported round its edge, so that there is a tendency for it to distort under its own weight. Yerkes Observatory, in America, has a 40-inch, and probably no refractors of larger aperture will be built. Mention must also be made of the 33-inch at Meudon Observatory, between Paris and Versailles. As I have used this telescope extensively for lunar work, I can well appreciate its capabilities.

This is not the place to discuss great observatories and vast instruments in any detail, so let us now return to the problems facing the amateur who is less lavishly equipped. He may have a small refractor or reflector, a pair of binoculars, a spectacle-lens home-made telescope or even nothing except his own eyes. Yet he can still take a real interest, and he can still make himself useful. To be a serious amateur observer you do not necessarily have to own a large and expensive telescope.

INTRODUCTION TO THE STARS

Few people can fail to be impressed with the glory of a brilliantly starlit night. It is sometimes hard to believe that instead of being able to see millions of stars one can never see more than a few thousand at any one moment. The whole sky gives the impression of being crowded, with here and there a particularly bright star outshining all its neighbours.

For convenience the stars are divided into grades or *magnitudes* of apparent brilliancy. The rule is the brighter the star, the smaller the magnitude. Stars of magnitude 6 are just visible to the naked eye under normal conditions; stars of magnitude 5 are brighter, 4 brighter still, and so on. Aldebaran, the reddish star in the constellation of the Bull, is very prominent, and is of about magnitude 1 (more accurately 0·8, so that it is slightly above the first magnitude). The four brightest stars in the sky have negative magnitudes, the most splendid of all, Sirius, being measured as $-1·4$. On this scale Venus, the brightest planet, is of magnitude $-4·4$, while the Sun is estimated at -26.

Any small telescope, or pair of binoculars, will extend the visible range below magnitude 6. A 2-inch refractor will reach down to magnitude 9·1; a 3-inch to 9·9; a 4-inch to 10·7; and a 6-inch to 11·6. The Palomar reflector can photograph stars with magnitudes as great as $+23$.

It is important to note that apparent magnitude is not necessarily a key to the star's real luminosity, for the simple reason that the stars are not all at the same distance from us. For example, Vega, the lovely blue star in the Lyre (magnitude 0·0), is very slightly brighter than Rigel in Orion (magnitude 0·1). Yet Vega, at a distance of only 26 light-years, is a mere 52 times as luminous as the Sun: the far more remote Rigel shines, as we have seen, as brilliantly as some 60,000 Suns. Another celestial searchlight is Deneb, in the Swan. To us it appears as the eighteenth brightest star in the sky, with a magnitude of 1·3, so that it is more than a magnitude fainter than Vega. Yet it is in fact 60,000 times more luminous than the Sun, and is over 1,500 light-years away.

Only two of the first-magnitude stars are within 10 light-years of us. One of these, Alpha Centauri (about 4 light-years), is too far south to be seen in Britain. The other is Sirius, which is 8½ light-years distant.

It is clear that the stars differ greatly in luminosity, but it would be misleading to suggest that the Sun is particularly feeble. It is, in fact, just about average, and many of our nearest stellar neighbours are dim in comparison. Proxima Centauri, for instance, has only 1/10,000 of the solar luminosity.

The stars are so remote that no telescope yet built will show them as anything except points of light. An instrument which shows a star as a disk is either faulty or out of adjustment; it is true to say that the smaller a star appears, the better the telescope and the observing conditions. Yet many of the stars are of great size. Betelgeux in Orion is known to be at least 250,000,000 miles in diameter, and even Betelgeux is by no means the supreme giant of the Galaxy.

As we can only see a star as a dot of light, it might be thought that our knowledge must necessarily be very limited. Fortunately, a great deal of information may be drawn from instruments based on the principle of the *spectroscope*.

We have seen that a glass prism will split up 'white' light into its component colours. In 1814 the German optician Fraunhofer investigated the spectrum of the Sun, and found that he observed a rainbow background crossed by dark lines. He went on to construct the first true astronomical spectroscopes, and it was found that the stars, too, produced similar effects, though not all stars yielded identical spectra.

In 1859, G. Kirchhoff explained the reason for the dark lines, and so laid the foundations of modern *astrophysics*. It is known that an incandescent solid or liquid, or gas under high pressure, produces a rainbow or *continuous* spectrum, while gas under low pressure will yield an *emission* spectrum made up of isolated bright lines. The essential point is that there are only 92 fundamental substances or *elements*,[1] and each of these elements produces its own particular lines—which cannot be duplicated by those of any other element. Sodium, for example, produces (among others) two bright yellow lines, so that when these are observed they at once reveal the presence of sodium vapour.

With a star, the bright gaseous surface yields a continuous spectrum. Overlaying the surface are layers of much more tenuous gases, which would normally give emission lines. What actually happens is that these gases have the effect of seeming to absorb some of the radiation from the background, and instead of appearing bright, the lines show up dark. In the spectrum of the Sun a double dark line is seen in the yellow region, and proves to be nothing more nor less than the familiar

[1] In recent years some more elements have been made artificially; but all these are unstable, and do not occur naturally on the Earth.

line due to sodium. We can thus tell that there is sodium in the Sun.

For convenience the stars are divided into various *spectral types*. Those of types B and A are white or bluish-white; F and G, yellowish; K, orange; and M, orange-red. In addition there are five more classes, each including a relatively small number of stars: W and O (white), and R, N and S (red). Each type is further subdivided, these divisions being indicated by figures from 0 to 9. For instance, Regulus in Leo is classed as B8.

The differences in colour of the stars indicate real differences in surface temperature. Spica in Virgo (type B) has a temperature of about 25,000° Centigrade; Sirius (type A) 11,000°; the Sun (type G) 6,000°; Betelgeux (type M) a mere 3,000°—though Betelgeux compensates for its lower temperature by its immense size, and has a luminosity 5,000 times greater than that of the Sun. Binoculars bring out the colours of the stars very strikingly.

At this point it may be of interest to list the fifteen brightest stars visible from Britain, giving some details of their spectral characteristics. These stars are generally regarded as being of the first magnitude, though the last four are actually fainter than magnitude 1·0. The remaining first-magnitude stars (Canopus, Alpha Centauri, Achernar, Agena and Acrux) are too far south to be seen in Britain.

In ancient times the stars were divided into groups, each of which was named—perhaps after a mythological god or hero (Orion, Perseus, Cepheus) or a living creature or common object (the Great Bear, the Lion, the Cup). Ptolemy, last of the great astronomers of classical times, who worked in Alexandria between A.D. 120 and 180

34

THE FIFTEEN BRIGHTEST STARS VISIBLE FROM BRITAIN

Name	Constellation	Magnitude	Spectrum	Colour	Distance, lt.-yrs.	Luminosity, Sun = 1
Sirius	Canis Major	−1·4	A0	White	8·6	26
Arcturus	Boötes	−0·1	K0	Orange	36	115
Vega	Lyra	0·0	A0	Bluish	26	55
Rigel	Orion	0·1	B8	White	900	60,000
Capella	Auriga	0·2	G0	Yellow	45	150
Procyon	Canis Minor	0·4	F5	Yellowish	11	7
Altair	Aquila	0·8	A5	White	16	10
Aldebaran	Taurus	0·8	K5	Orange	68	165
Betelgeux	Orion	variable	M0	Reddish	520	15,000
Antares	Scorpio	0·9	M0	Red	520	9,600
Spica	Virgo	1·0	B2	White	220	1,800
Fomalhaut	Piscis Australis	1·1	A3	White	23	23
Pollux	Gemini	1·2	K0	Orange	35	34
Deneb	Cygnus	1·3	A2	Yellowish	1,600	60,000
Regulus	Leo	1·3	B8	White	84	170

and produced a great book known to us by its Arab title of the *Almagest*, drew up a star catalogue and enumerated 48 constellations; the list has since been extended. Very few of the groups have outlines which bear the slightest resemblance to the objects after which they are named, but the system has stood the test of time, and will certainly never be altered now.

The names are used in their Latin forms, and it is best to keep to these; thus the Great Bear becomes 'Ursa Major' and the Lion 'Leo'. A list of the constellations visible in Britain is given below. Unfortunately some of the really interesting groups lie well to the south, and never rise in our latitudes, while others (such as Argo Navis, the Ship) show only small parts of themselves above the British horizon. In other cases, notably with Eridanus and Scorpio, parts of the constellation are too far south to be seen.

CONSTELLATIONS VISIBLE FROM BRITAIN

Latin name	English name
Andromeda	Andromeda
Aquarius	The Water-bearer
Aquila	The Eagle
Aries	The Ram
Auriga	The Charioteer
Boötes	The Herdsman
Camelopardus	The Giraffe
Cancer	The Crab
Canes Venatici	The Hunting Dogs
Canis Major	The Great Dog
Canis Minor	The Little Dog
Capricornus	The Sea-Goat
Cassiopeia	Cassiopeia

36

Latin name	English name
Cepheus	Cepheus
Cetus	The Whale
Coma Berenices	Berenice's Hair
Corona Borealis	The Northern Crown
Corvus	The Crow
Crater	The Cup
Cygnus	The Swan
Delphinus	The Dolphin
Draco	The Dragon
Equuleus	The Little Horse
Eridanus	The River
Gemini	The Twins
Hercules	Hercules
Hydra	The Sea-Serpent
Lacerta	The Lizard
Leo	The Lion
Leo Minor	The Little Lion
Lepus	The Hare
Libra	The Scales
Lynx	The Lynx
Lyra	The Lyre
Monoceros	The Unicorn
Ophiuchus	The Serpent-Bearer
Orion	Orion
Pegasus	The Flying Horse
Perseus	Perseus
Pisces	The Fishes
Piscis Australis	The Southern Fish
Sagitta	The Arrow
Sagittarius	The Archer
Scorpio	The Scorpion
Sculptor	The Sculptor
Scutum Sobieskii	Sobieski's Shield
Serpens	The Serpent
Sextans	The Sextant
Taurus	The Bull

Latin name	English name
Triangulum	The Triangle
Ursa Major	The Great Bear
Ursa Minor	The Little Bear
Virgo	The Virgin
Vulpecula	The Fox

In 1603, the German astronomer Bayer published a star atlas in which he allotted the stars Greek letters according to their constellations and apparent magnitudes. The system was clearly a good one; most of the really bright stars have individual names, but it would be both cumbersome and difficult to attempt to find a name for each separate star. Bayer's idea was to designate the brightest star in a constellation Alpha, the second Beta, and so on; thus Sirius, the brightest star in Canis Major, would become Alpha Canis Majoris. In some cases the letters are out of order, as in Orion, where Beta (Rigel) is brighter than Alpha (Betelgeux). This does not much matter, and Bayer's letters are still used. The Greek alphabet is as follows:

α	Alpha	η	Eta	ν	Nu	τ	Tau
β	Beta	θ	Theta	ξ	Xi	υ	Upsilon
γ	Gamma	ι	Iota	o	Omicron	ϕ	Phi
δ	Delta	κ	Kappa	π	Pi	χ	Chi
ϵ	Epsilon	λ	Lambda	ρ	Rho	ψ	Psi
ζ	Zeta	μ	Mu	σ	Sigma	ω	Omega

In the following chapters, sufficient instructions are (it is hoped) given to enable the beginner to find his way among the constellations, and to identify the various groups. Here, as in all things, a little practice will work wonders, and after a few weeks it should prove easy to recognize the stars almost at a glance.

In addition to being of different colours, the stars have their own special characteristics, and there is endless variety in the stellar heavens. In particular it is worth noting the double stars, the variables and the clusters and nebulæ.

Look closely at Mizar (ζ Ursæ Majoris), the second star in the tail of the Great Bear or handle of the Plough, and you will see that it has a much smaller star, Alcor, close beside it. Use a small telescope, and it will be evident that Mizar itself is double, and is made up of two separate stars, one rather brighter than the other. The two are so close together that to the naked eye they appear as a single point of light.

This is no mere chance lining-up. The components of Mizar are physically associated, and form a *binary* system. The two are revolving round their common centre of gravity, though their relative movement is so slow that it cannot be detected except over periods of many years.

Binary stars prove to be surprisingly common in the sky. Mizar is a wide system, and easily divided; so too is γ Virginis, not far from Spica, though it is less striking now than it used to be. Others are much closer, and require more powerful instruments. Capella, for instance, is an excessively close binary, separable only with the aid of the world's greatest telescopes.

The components of Mizar are not very unequal, but there are many cases of a bright star being accompanied by a much fainter companion. Undoubtedly the most striking example is Albireo or β Cygni, in the Swan, which is made up of a yellowish third-magnitude star attended by a fifth-magnitude companion which some observers describe as green and others as bluish. Antares, which is strongly red, has a green attendant of magnitude 7; Polaris is accompanied by a ninth-magnitude companion which is not difficult to see with a 3-inch refractor.

Even more interesting are the multiple stars, of which the best example is ε Lyræ, close to Vega. Keen eyes can see that it is made up of two components, each of about the fourth magnitude. A 3-inch refractor shows that each component is itself double, so that we have a double-double or quadruple system. Another famous multiple is θ Orionis, known commonly as 'the Trapezium' for reasons which will be obvious to anyone who has looked at it through a telescope of moderate aperture.

We also meet with clusters of stars. The most famous of these is the Pleiades or Seven Sisters, which lies in Taurus. To the naked eye it appears as a closely knit group containing between six and a dozen stars; binoculars show many more, and altogether it is known that the cluster includes at least 200 members. Other 'open clusters' of this type are the Hyades, round Aldebaran, and Præsepe or the Beehive, in Cancer. Quite different are the globular clusters, which have the appearance of 'balls of stars', closely packed toward the centre of the system. Most of these are faint objects, and it is a pity that the two brightest, ω Centauri and 47 Tucanæ in the Toucan—47 is a catalogue number—never rise in Britain. However, we have one splendid globular available, the great cluster in Hercules, which is just visible to the naked eye.

Other fascinating objects are the variable stars, which alter in brilliancy over short periods. Betelgeux in Orion is one such object. Sometimes it far outshines Aldebaran, and has even been known to rival Rigel; at others it is little brighter than Aldebaran. The changes are real, and are due to alterations in the diameter of the star.

Betelgeux has no definite period, and its behaviour cannot be predicted with any accuracy, but other variables are as regular as clockwork. δ Cephei, a relatively inconspicuous star in the nothern sky, changes between magnitudes 3·7 and 4·3 in a period of 5·37 days, and its magnitude for any moment may be forecast years in advance. Less reliable are the long-period variables, of which Mira (o Ceti) in the Whale is the most famous. At maximum Mira may attain the second magnitude; at minimum it sinks to below 9, so that even binoculars will not show it. The period is about 331 days, but is not constant, and may alter within relatively narrow limits.

It is worth noting that many of the long-period and some of the irregular variables are reddish, with spectra of type M, and so may be easily recognized. Unfortunately only a few of them are bright, since all are comparatively remote.

Look below the three bright star which make up the Belt of Orion, and you will see what looks like a misty patch. Binoculars bring out the appearance even more clearly, and the impression is one of luminous gas. This is indeed a true picture; the patch is a *galactic nebula*, composed chiefly of very tenuous hydrogen, and shines because there are stars contained in it. Many other such nebulæ are known, but the Sword of Orion must take pride of place.

In 1781 the French astronomer Charles Messier drew up a catalogue of the most conspicuous clusters and nebulæ, and numbered them. His numbers are still used, so that, for instance, Præsepe is M.44 and the Orion nebula M.42.

Messier also included some objects which he regarded as nebulæ, but which are now known to be separate galaxies far beyond the limits of the Milky Way. The Andromeda Galaxy, dimly visible without a telescope on a clear night, is M.31 in his catalogue.

In astronomy, as in most other sciences, the unexpected sometimes happens, and now and then a brilliant star will flare up in a position where no bright star has been present before. Such an appearance is known as a *nova*, or new star. The name is rather misleading, since a nova is not properly 'new' at all. What happens is that a previously very faint star suffers an internal outburst which causes a tremendous, though temporary, increase in the output of radiation. After remaining prominent for a few days or weeks the nova sinks back to its former obscurity. Among the bright novæ of the present century were the stars in Perseus (1901), Aquila (1918), Hercules (1934) and Cygnus (1975). Amateurs have been responsible for the detection of some of these novæ; for example, the interesting Nova Herculis 1934 was found by a British meteor observer, J. P. M. Prentice. Yet a word of warning may be timely here. If you see a bright object which is not to be found on your star-maps, check carefully to make sure that it is not a planet. 'Novæ' reported during the past few years include such old friends as Saturn, Jupiter and Mars!

The most successful British nova-hunter is G. E. D. Alcock, who is by profession a schoolmaster in Peterborough. In 1967, using his specially mounted binoculars, he found a naked-eye nova in Delphinus, which proved to be of exceptional interest; it faded very slowly, and in mid-1982 was still only slightly below the 12th magnitude. In the following year Alcock found a 5th-magnitude nova in Vulpecula, which faded more quickly; and in 1970 he completed a trio of triumphs by his discovery of a rather fainter nova in Scutum. This shows what can be done by someone who is really skilful and dedicated. Alcock has memorized the positions and magnitudes of over 30,000 stars, so that he can recognize a newcomer instantly.

To the owner of a powerful telescope, the night sky offers endless scope. Doubles, variables, clusters and nebulæ may be seen in their thousands and useful work may be carried out—in checking the fluctuations of the irregular and long-period variables, for example. In the following pages, however, attention will be concentrated on objects which may be seen either with the naked eye or with binoculars or very small telescopes. Once these have been mastered, our would-be astronomer will be ready to turn to a more detailed set of star-maps.

CIRCUMPOLAR STARS

The best method of finding one's way around an unfamiliar city is to note a few prominent objects and use them as landmarks. In the sky, too, we have certain very conspicuous features which may be used as guides. So far as we in Britain are concerned, the most valuable of all is the Great Bear, because it is extremely easy to find and never drops below the horizon. In winter we have an even more brilliant guide-group in Orion, the Hunter.

In the following pages, the various constellations will be described one by one. Stars of magnitude 3·0 or brighter are listed, together with their spectral types and distances in light-years. The 'absolute magnitude' is also given. This is a measure of a star's real luminosity, and represents the magnitude which a star would appear were it at a standard distance of 32·6 light-years. Thus the Sun would show as a dim star of magnitude 4·8, and would be very inconspicuous indeed; on the other hand Bellatrix in Orion would shine of magnitude −4·2, which is about the brilliancy of the planet Venus in our skies, while others would be even brighter. It must, of course, be realized that for remote stars the distances and luminosities are decidedly uncertain, and different authorities give different values.

For the moment, then, let us begin with the circumpolar groups, which are always to be seen from Britain whenever the skies are dark and clear. We will leave the stars of the far south until last, since to Britons they never rise above the horizon.

URSA MAJOR The Great Bear

Ursa Major is one of the oldest of the constellations, and is included in the forty-eight listed by Ptolemy. In mythology, Ursa Major was originally Callisto, attendant to the goddess Juno and daughter of King Lycaon of Arcadia. Her beauty surpassed Juno's own, and the jealous goddess was enraged as a result. To protect Callisto, Jupiter, king of Olympus, turned her into a bear. Unfortunately Callisto's son, Arcas, saw the bear while he was out hunting, and was about to kill it with his spear when Jupiter intervened.

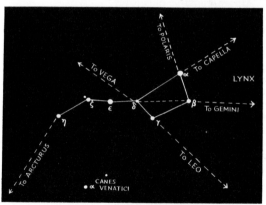

CHIEF STARS

		Visual Magnitude	Spectrum	Absolute Magnitude	Distance lt.-yrs.
ε	Alioth	1·79	A0	+0·2	68
α	Dubhe	1·81	K0	−0·7	107
η	Alkaid	1·87	B3	−2·1	210
ζ	Mizar	2·06	A2	0·1	88
β	Merak	2·37	A1	0·5	78
γ	Phad	2·44	A0	0·2	90

45

He turned Arcas into a bear also, and placed both animals in the sky.

These six stars, together with a seventh, δ or Megrez (magnitude 3·3), make up the famous Plough, also known as King Charles' Wain and—in America—as the Big Dipper. It is so conspicuous that it cannot be overlooked. During spring evenings it is almost overhead; even when at its lowest, as during winter evenings, it is still well above the northern horizon. As the chart shows, it can be used to find many other stars: Arcturus in Boötes, Regulus in Leo and Capella in Auriga, to name only three.

Dubhe is a yellowish star, and binoculars will show the difference in hue between it and its companions. Mizar forms a naked-eye double with Alcor (magnitude 5) and, as we have seen, a low power on a small telescope reveals that Mizar is itself double. It is without doubt one of the most splendid binaries in the heavens. Megrez is a magnitude fainter than any of the other Plough stars, but according to Ptolemy and other astronomers of ancient times it used to be equal to the rest. There is a definite possibility that it has faded during the past 2,000 years, and it may therefore be a peculiar kind of irregular variable.

URSA MINOR The Little Bear

One of Ptolemy's forty-eight constellations. In mythology, the Bear is the 'Arcas' of the Callisto legend. The group is notable mainly because it contains the north celestial pole, which lies less than one degree from Polaris.

Ursa Minor is easy to find. Two of the Plough stars, Merak and Dubhe, are known as the Pointers because they point directly to Polaris; moreover, Polaris seems to remain almost motion-

46

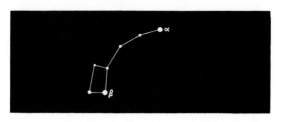

CHIEF STARS

		Visual Magnitude	Spectrum	Absolute Magnitude	Distance lt.-yrs.
α	Polaris	1·99	F8	−4·6	680
β	Kocab	2·04	K4	−0·5	105

less in the sky. The group is shaped a little like Ursa Major, but apart from Polaris and Kocab its stars are much less bright.

Polaris has a ninth-magnitude companion, easy to see with a 3-inch refractor, but difficult with a 2-inch. It is worth using binoculars, or a low power, to look first at Polaris, and then at Kocab. The strong orange colour of Kocab, the so-called 'Guardian of the Pole', will be immediately obvious.

CAMELOPARDUS The Giraffe

Camelopardus is not one of the ancient constellations; it is found on the star-map drawn by Hevelius of Danzig in 1690, though it may have been introduced rather earlier. Some historians suppose that it is meant to represent the camel which carried Rebecca to Isaac.

The constellation lies more or less between Ursa Major and the prominent W of stars which makes up Cassiopeia. It contains no stars as bright as magnitude 4, and neither are any interesting objects visible with small telescopes. All things considered, Camelopardus marks one of the most barren regions of the entire sky.

CANES VENATICI The Hunting Dogs

This is another of the constellations formed by Hevelius in 1690. There are two dogs, Asterion and Chara, meant to be held in leash by the herdsman (Boötes) as they chase the two Bears round the celestial pole.

CHIEF STAR

	Mag.	Spec.	Dist., lt.-yrs.	Luminosity, Sun=1
α Cor Caroli	2·9	A0	91	52

The only bright star, Cor Caroli ('Charles' Heart', named by Edmond Halley, second Astronomer Royal, in memory of Charles I), is easy to recognize, as it lies below the tail of the Great Bear, and there are no conspicuous stars close to it. An extra pointer may be obtained by using Polaris and Alioth as direction-finders. Cor Caroli has a companion of magnitude 5·6, easy to see with a low power, since the separation is rather wide. This is not a binary system, but an 'optical double'; the companion is much more remote than Cor Caroli itself, and simply happens to lie in almost the same line of sight as seen from the Earth. There are no other objects in the constellation of interest except to owners of powerful telescopes. Canes Venatici is included with the map showing Ursa Major (page 45).

CASSIOPEIA

Cassiopeia is one of Ptolemy's constellations, and is among the leading groups of the northern hemisphere.

In ancient mythology, Cassiopeia was a proud queen, wife of King Cepheus and mother of the lovely princess Andromeda. Cassiopeia was unwise enough to boast that Andromeda's beauty was greater than that of the sea-nymphs or Nereids. This angered Neptune, the sea-god, who sent a monster to ravage Cepheus' land. The king and queen were in despair and consulted the Oracle. What they heard made them lament even more loudly; Andromeda was to be chained to a rock by the shore and left as a prey for the monster. However, this is one old legend which has a happy ending; in the proverbial nick of time, Andromeda was saved by the intervention of the gallant hero Perseus.

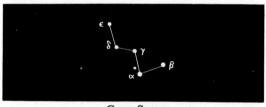

CHIEF STARS

		Visual Magnitude	Spectrum	Absolute Magnitude	Distance, lt.-yrs.
α	Shedir	var.	K0	−1·1	150
β	Chaph	2·26	F2	1·6	45
γ	Tsih	var.	B0	−0·3	96
δ	Ruchbah	2·67	A5	2·1	43

Together with ε (magnitude 3·4) these stars make up the familiar W. Alioth and Polaris act as pointers to it, but it is so distinctive that it will be recognized without the slightest difficulty.

Shedir is a reddish irregular variable, usually slightly brighter than its companion; its average magnitude is 2·2. γ is an extraordinary star, fluctuating between magnitudes $1\frac{1}{2}$ and $3\frac{1}{4}$. For the last few years it has remained around 2·3, so that it has been about equal to β. The observer who compares it with β for several consecutive nights may notice some changes in the relative brilliancy of the two, but it is only right to add that for long periods both it and Shedir remain virtually steady. Indeed, the Russian astronomer Kukarkin does not consider that Shedir is variable at all.

Shedir has a ninth-magnitude companion, difficult to see in binoculars but easy with a small telescope. This is another optical double, and not a binary system.

The Milky Way runs through Cassiopeia, and glorious star-fields may be found by sweeping with binoculars or a low power.

CEPHEUS

One of Ptolemy's constellations. In mythology, Cepheus was the husband of Cassiopeia and father of Andromeda.

Cepheus is by no means so conspicuous as his wife. Part of the constellation lies between Cassiopeia and Polaris; the best guide is to use two of the W stars in Cassoipeia (Shedir and γ) as pointers.

There are, however, two extremely interesting variable stars. δ Cephei forms a small triangle with its neighbours ζ (3·6) and ϵ (4·2). Whereas ζ and ϵ are constant, δ varies between magnitudes 3·7 and 4·3 in a period of 5·37 days. The period is absolutely constant, and the star is the prototype of the class of variables known as *Cepheids*. Its fluctuations may be easily noted from night to night.

Not far off is a much fainter variable, μ. This time the spectrum is of type M, and the magnitude changes irregularly between 4 and 6, so that when

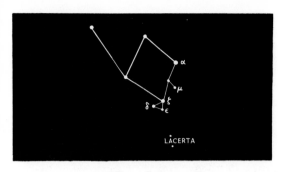

LACERTA

	Visual Magnitude	Spectrum	Absolute Magnitude	Distqnce, lt.-yrs.
α Alderamin	2·44	A7	1·4	52

at its faintest the star is barely visible to the naked eye. Yet binoculars will suffice to bring out its glorious red hue. Sir William Herschel, a famous observer who lived a century and a half ago, called it 'the Garnet Star', and it is certainly well worth looking at.

DRACO The Dragon

An original constellation. Mythologically it is often identified with Ladon, the watchful dragon which guarded the golden apples in the Garden of the Hesperides, and was finally killed by Hercules. It has also been regarded as the guardian dragon of a sacred spring, killed by the hero Cadmus, who afterwards planted the dragon's teeth and raised a whole army of armed men—who immediately fought amongst themselves and slew each other, leaving only the five survivors who helped Cadmus to found the city of Bœotia.

51

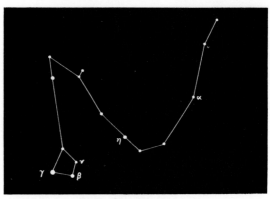

CHIEF STARS

		Visual Magnitude	Spectrum	Absolute Magnitude	Distance, lt.-yrs.
γ	Eltamin	2·21	K5	−0·4	108
η	Aldhibain	2·71	G8	0·3	103
β	Alwaid	2·77	G2	−2·1	310

Draco is not a prominent group. It is long and winding; two of its brightest stars, γ and β, lie not far from the brilliant Vega in Lyra, and the constellation then curls in a line of inconspicuous stars round Ursa Minor, ending almost between Dubhe and Polaris. Thuban or α Draconis, magnitude 3·6, used to be the north polar star in ancient times. Since then effects of precession have carried the pole away from Thuban and almost to Polaris.

ν, in the dragon's 'head', is a wide double. Each component is of magnitude 4·5, and the pair are well separated in binoculars.

LACERTA The Lizard

Lacerta, first found on Hervelius' star map of

52

1690, is a small constellation near Cepheus. It contains no star brighter than the fourth magnitude, and no interesting objects for small telescopes, though in 1936 a bright nova appeared in it and attained the second magnitude before fading away to obscurity. Lacerta is shown in the map on page 51 with Cepheus.

LYNX The Lynx

Yet another of Hevelius' additions. It lies between the Pointers (Dubhe and Merak) on the one side, and Capella and the Twins (Castor and Pollux) on the other. Lynx is decidedly barren, and contains no star as bright as the third magnitude. Like Lacerta it seems to have no mythological associations. Most of it is circumpolar, and it is shown in the map on page 45 with Ursa Major.

It will be seen that of all the northernmost constellations only two—Ursa Major and Cassiopeia—are really prominent. Actually three of the first-magnitude stars (Capella, Vega and Deneb) are also circumpolar, but all three go down almost to the horizon, and are best treated with the 'seasonal' groups. So let us now carry out a brief survey of the heavens from month to month, beginning with the sky as it is seen during a winter evening.

WINTER STARS

Winter is certainly the best time to start learning star recognition. In addition to the Great Bear, which is always useful, we have the second of our main 'direction-finders', Orion—the Hunter with his brilliant retinue, which includes Sirius, the brightest star in the whole sky.

On a winter evening Orion may be seen in the south, his pattern quite unmistakable both because of its distinctive shape and because of the brilliance of its stars. From it we may find not only Sirius but also many important groups such as Taurus (the Bull), Gemini (the Twins) and Auriga (the Charioteer or Wagoner). Capella, in Auriga, is almost directly overhead.

The Great Bear lies in the north-east, and Regulus in the Lion is well above the eastern horizon. The W of Cassiopeia is high up rather to the west of the *zenith* or overhead point, and the Square of Pegasus may be seen in the west. Vega is very low down, almost at the northern horizon, and not far off lies the cross of Cygnus, with the bright star Deneb. The Milky Way extends in a glorious arc from Cygnus through Cassiopeia and Auriga, passing between Orion and the Twins and extending down to the southern horizon. The south-west area is occupied by two large rather ill-defined groups: Eridanus (the River), whose brightest stars never rise in Britain, and Cetus (the Whale), which is better placed during autumn evenings.

ORION The Hunter

Orion is one of Ptolemy's constellations, and

represents the mythological hunter. Many legends are associated with him. According to one account he was the son of Neptune and Euryale, and boasted that he could conquer any animal which earth could produce—whereupon Juno, who was jealous of him, caused a giant scorpion to appear out of the ground and bite Orion in the foot, killing him. Subsequently the pleadings of Diana,

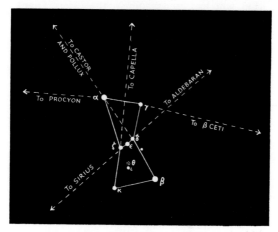

CHIEF STARS

		Visual Magnitude	Spectrum	Absolute Magnitude	Distance, lt.-yrs.
β	Rigel	0·08	B8	−7·1	900
α	Betelgeux	var.	M2	−5·6	520
γ	Bellatrix	1·64	B2	−4·2	470
ε	Alnilam	1·70	B0	−6·8	1,600
ζ	Alnitak	1·79	O9·5	−6·6	1,600
κ	Saiph	2·06	B0·5	−6·9	2,100
δ	Mintaka	var.	O9·5	−6·1	1,500
ι		2·76	O9	−6·1	2,000

the goddess of hunting, led to his being placed in the heavens directly opposite to the scorpion (Scorpio) so that he should suffer no more harm from it. Another legend relates how Diana was in love with Orion, and thus angered Apollo, who persuaded her to try her skill at archery by shooting at a certain object in the sea. She aimed, and hit the distant mark—which proved to be the head of Orion, who had been wading in the waters. Diana's arrow having killed him, the goddess placed him among the stars.

Orion is truly magnificent. Two of its stars, Rigel and Betelgeux, are of the first magnitude; Rigel is almost pure white while Betelgeux is a glorious orange–red and is a fine sight in binoculars. Betelgeux has a diameter of some 250,000,000 miles, so that its vast globe could contain the entire orbit of the Earth round the Sun. It is an irregular variable, with a magnitude range of between 0·1 and 0·9; on rare occasions it has been known to rival Rigel. but is more generally of around magnitude 0·5. There is a rough period of about 5 years, but the fluctuations cannot be predicted with any accuracy. Its changes are due to real alterations in its diameter, and hence in its output of radiation.

Apart from Betelgeux, all the leading stars of Orion are very hot and white. Particularly notable are the three members of the Hunter's Belt (Alnilam, Alnitak and Mintaka). Mintaka is slightly variable; its average magnitude is 2·5, and it has a seventh-magnitude companion. A 3-inch refractor will also show the seventh-magnitude attendant of Rigel.

Below the Belt lies the Hunter's Sword. To the naked eye it appears as a faint misty patch; binoculars or a small telescope reveal the presence of shining gas—the brightest of the galactic nebulæ, No. 42 in Messier's catalogue. Immersed in nebulosity are many stars, including the famous 'Trapezium', θ Orionis, whose four chief components are not difficult to see with a 3-inch telescope. According to modern theory, nebulæ of this type are the birthplace of stars, and it is very likely that fresh stars are being produced inside the Orion nebula, though of course the process is a very slow one judged by terrestrial standards.

As a direction-finder Orion is unrivalled, as will be seen from the diagram given on page 55.

AURIGA The Charioteer

An original group. In mythology Auriga represents Erechthonius, son of Vulcan, who was born deformed, and was reared by Minerva with-

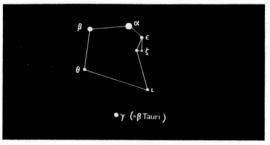

CHIEF STARS

		Visual Magnitude	Spectrum	Absolute Magnitude	Distance, lt.-yrs.
α	Capella	0·05	G8	−0·6	45
β	Menkarlina	1·86	A2	−0·3	88
ι		2·64	K3	−2·4	330
θ		2·65	B9·5	0·1	108
ε		var.	F0	−7·1	3,400

57

out the knowledge of the other Olympians. When he reached manhood he became King of Athens, and invented the four-horse chariot, for which Jupiter rewarded him by placing him in the sky.

In older maps the star Al Nath was included in this constellation as γ Aurigæ, but has now been transferred to Taurus and is known as β Tauri.

Capella is always easy to identify. It is yellowish in colour, since its spectrum resembles that of the Sun, and it is very high up during winter evenings; it may indeed pass over the zenith. It is a binary, but the two components are so close together that giant telescopes are needed to separate them. The pattern of Auriga bears some resemblance to a kite, extending from Capella in the general direction of Orion.

Close to Capella in the sky are three fainter stars forming a triangle, the Hædi or Kids. Two of these, ϵ and ζ, are most interesting, since they are binaries of a special type; every 127 years the bright component of ϵ is eclipsed by an invisible companion of uncertain nature, and fades down by a magnitude. The most recent eclipse was that of 1902–3.

CANIS MAJOR The Great Dog

One of Ptolemy's constellations. It represents the larger of Orion's dogs, attending his master faithfully during the never-ending journeys across the sky.

Sirius is unmistakable because of its brilliancy, which far exceeds that of any other star. The three members of Orion's Belt point almost directly to it. Apart from α Centauri, which never rises in Britain, Sirius is the closest of the really bright stars; in reality it is not outstanding, even though it is 26 times as luminous as the Sun. It has a small companion, which is not visible except with large

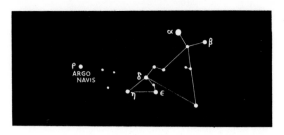

CHIEF STARS

		Visual Magnitude	Spectrum	Absolute Magnitude	Distance, lt.-yrs.
α	Sirius	−1·43	A1	1·45	9
ε	Adara	1·48	B2	−5·1	680
δ	Wezea	1·85	F8	−7·1	2,100
β	Mirzam	1·96 (v)	B1	−4·8	750
η	Aludra	2·46	B5	−7·1	2,700

telescopes. Sirius is known popularly as the Dog-Star, and its glittering whiteness makes it a splendid object in binoculars. Apart from Sirius there are few notable objects in Canis Major.

CANIS MINOR The Little Dog

Another of Ptolemy's groups, representing the junior of Orion's two hounds.

59

		Visual Magnitude	Spectrum	Absolute Magnitude	Distance, lt.-yrs.
α	Procyon	0·37	F5	2·7	11
β	Gomeisa	2·91	B7	−1·1	210

Procyon is conspicuous because of its brightness; like Sirius it has a faint companion visible only in large telescopes. β is the only other prominent star in Canis Minor.

ERIDANUS The River

An original constellation. According to fable it represents the Italian river now known as the Po. The story relates how the reckless youth Phæthon obtained permission to drive the sun-chariot across the sky for one day; how the horses bolted; and how Jupiter, in order to save the world from destruction, struck Phæthon with a thunderbolt, toppling him from the chariot and sending him headlong into the river below.

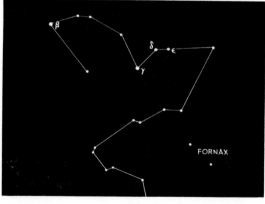

	Visual Magnitude	Spectrum	Absolute Magnitude	Distance, lt.-yrs.
β Kursa	2·79	A3	0·9	78

Only the northern part of Eridanus is visible from Britain. The rest, including the first-magnitude star Achernar, never rises in our latitudes. Of the part we can see the only brightish star is β, which lies not far from Rigel. The constellation occupies a large area to the west of Orion, but contains no objects of note from the viewpoint of the amateur observer.

GEMINI The Twins

Another of Ptolemy's original forty-eight groups, and one of the constellations of the Zodiac. It takes its name from Castor and Pollux, two

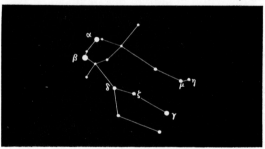

CHIEF STARS

	Visual Magnitude	Spectrum	Absolute Magnitude	Distance, lt.-yrs.
β Pollux	1·16	K0	1·0	35
α Castor	1·62	A1 and A5	1·3 and 2·3	45
γ Alhena	1·93	A0	−0·6	105
μ Tejat	2·92 (v)	M3	−0·6	160
ε Mebsuta	3·00	G8	−4·6	1,080

61

of the mythological heroes—twin boys, sons of a Spartan king, Tyndarus, and his queen Leda. Pollux was immortal, while Castor was not. When Castor was killed, Pollux was so grief-stricken that he pleaded to be allowed to share his immortality with his brother, whereupon Jupiter placed both youths in the sky.

Gemini is one of the grandest of the northern constellations, particularly as the Milky Way passes through it. Rigel and Betelgeux act as pointers to the two twins; another way to find them is by using Megrez and Merak in the Great Bear. Pollux is decidedly brighter than Castor, though apparently this was not the case in ancient times. Moreover the colours of the two differ, as binoculars will show: Pollux is orange, Castor white. The rest of the constellation extends from the two Twins in the direction of Orion; Alhena, which is bright enough to be conspicuous, lies roughly between Pollux and Betelgeux.

Castor is a splendid binary, and may be separated with a moderate telescope. The components revolve round their common centre of gravity in a period of 350 years. The third-magnitude star δ has an eighth-magnitude companion, and there are also two interesting variables: ζ, which changes regularly between magnitudes 3·7 and 4·3 in a period of just over 10 days, and η, which has a range of between 3·3 and 4·2 and a period of 231 days. η is reddish, and lies close to another reddish star, the third-magnitude Tejat. Its changes in light are slow, but are not hard to detect over periods of several weeks. Near by is a fine star-cluster, Messier 35, just visible to the naked eye and conspicuous in binoculars.

Gemini is a very rich constellation, and is worth sweeping with a low power.

LEPUS The Hare

Lepus is by no means prominent, but is one of Ptolemy's forty-eight groups. According to legend Orion was particularly fond of hunting the hare, and so a hare was placed close to him in the sky.

CHIEF STARS

	Visual Magnitude	Spectrum	Absolute Magnitude	Distance, lt.-yrs.
α Arneb	2·58	F0	−4·6	900
β Nihal	2·81	G5	0·1	113

Lepus lies below Orion. It contains some interesting objects, but none which is of interest to the user of binoculars or a small telescope—apart from R Leporis, a long-period variable, which is intensely red (it has been nicknamed the Crimson Star) and is just visible to the naked eye when at maximum. It has a period of 430 days.

MONOCEROS The Unicorn

Monoceros is included in Hevelius' map of 1690, but in all probability it was formed rather earlier. It represents the fabled unicorn, or 'horned horse', but there are no mythological legends attached to it.

The constellation lies mainly in the area bounded by Sirius, Procyon, the Twins and Betelgeux. It is worth sweeping with binoculars, since the Milky Way runs through it; and round the faint star 12

Monocerotis there is an open cluster just visible to the naked eye.

PERSEUS

An original constellation. In mythology, Perseus was the gallant hero who rescued Andromeda from the sea-monster while he was returning from his expedition against the terrible Gorgon, Medusa.

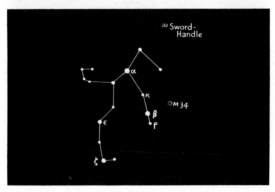

CHIEF STARS

		Visual Magnitude	Spectrum	Absolute Magnitude	Distance, lt.-yrs.
α	Mirphak	1·80	F5	−4·4	570
β	Algol	2·06-3·28	B8	−0·5	105
ζ		2·83	B1	−6·1	1,000
ε		2·88	B0·5	−3·7	680
γ		2·91	G8	0·3	113

Perseus contains no first-magnitude star, but is nevertheless a splendid constellation. It lies not far from Capella, but the best pointer to it is provided by two of the stars in the W of Cassiopeia. The Milky Way flows through it, and the area is rich in superb star-fields, as binoculars will show. There are two clusters visible to the naked eye. One, Messier 34, is a typical loose or open cluster. The other, the famous Sword-Handle (not to be confused with the Sword of Orion), is shown by any small telescope to be made up of two separate clusters close together in the sky. Oddly enough, Messier did not include them in his catalogue.

Algol, or β Persei, is an interesting star. Normally it shines of magnitude 2·3, almost equal to Polaris, but periodically it seems to 'wink' slowly, fading for 5 hours until its magnitude is only $3\frac{1}{2}$. After a short minimum of 20 minutes or so it slowly regains its lustre. Actually Algol is not truly variable at all; it is associated with a fainter binary companion which periodically passes in front of it and blocks out much of the brighter component's light. Algol is the prototype of these *eclipsing binaries*, of which many are known. An old name for it was 'the Demon', which is appropriate enough, since it marks the head of Medusa—though apparently its changes in light were not noticed until the seventeenth century. The times of minima are predicted in various yearly almanacs, and the fluctuations are easy to follow with the naked eye.

ρ Persei, not far from Algol, is a reddish irregular variable of about the fourth magnitude. κ (magnitude 4·0) makes a good comparison star. The changes are, however, slow and by no means obvious.

TAURUS The Bull

One of the Zodiacal constellations, included in Ptolemy's list. According to legend Jupiter, the king of the gods, fell in love with Europa, daughter of King Agenor of Crete, and decided to abduct her. He therefore assumed the form of a white bull

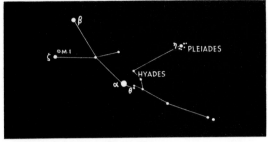

	Visual Magnitude	Spectrum	Absolute Magnitude	Distance, lt.-yrs.
α Aldebaran	0·86 (v)	K5	−0·7	68
β Alnath	1·65	B7	−3·2	300
η Alcyone	2·86	B7	−3·2	541
ζ	3·07	B2	−4·2	940

and induced Europa to ride on his back—whereupon Jupiter dashed into the sea and swam away with her. The story is charmingly told by Nathaniel Hawthorne in his immortal *Tanglewood Tales*.

Taurus is easy to find; its chief star, Aldebaran, is lined up with Orion's Belt, and is identifiable both by its brightness and by its strong reddish hue, which resembles that of Betelgeux.

The constellation abounds in spectacular objects, the most famous being the star-cluster known as the Pleiades or Seven Sisters, surrounding the third-magnitude star Alcyone (η Tauri). With the naked eye at least seven Pleiads may be seen on a clear night, while keen-sighted persons will count more; the record is said to be nineteen. Binoculars will show further members of the group. Telescopically it is best to use a very low power in order to see all the cluster in one field.

The Hyades, round Aldebaran, are more scattered, and are best viewed with binoculars. Oddly enough Aldebaran is not a true member of the cluster; it is much closer to us, and simply happens to lie in the same direction as seen from Earth. One of the Hyads, θ Tauri, is a naked-eye double.

The Crab Nebula, not far from ζ, is a fascinating object; it represents the wreck of a *supernova*, or stellar explosion, witnessed by Chinese observers in the year 1054. Large instruments show it as an extensive mass of gas, but in small telescopes it is too faint to be properly seen. Some observers state that it is visible in binoculars, but I confess that I have never been able to see it with anything less than a 3-inch refractor.

Supernovæ, unlike ordinary novæ, are very rare. Since 1054 only two have appeared in our own Galaxy; one in Cassiopeia (1572) and the other in Ophiuchus (1604). At maximum a supernova may emit as much light as 200,000,000 or even 300,000,000 Suns, so that it is a real stellar catastrophe during which the exploding star blows much of its material away into space. The gas which makes up the Crab Nebula is still expanding outward from the old explosion-centre. (For *pulsar* in the Crab, see page 130.)

FORNAX The Furnace

Fornax—known originally as Fornax Chemica, the Chemical Furnace—was introduced by the French astronomer Lacaille in 1752. It lies near Eridanus, and contains no bright stars or objects of interest. It is included in the map with Eridanus (page 60).

COLUMBA The Dove

A constellation introduced by Royer in 1679. It is

said to represent the dove which Noah released from the Ark.

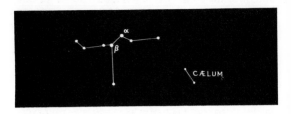

CHIEF STAR

	Visual Magnitude	Spectrum	Absolute Magnitude	Distance, lt.-yrs.
α Phakt	2·64	B8	−0·6	140

Columba is always too low to be well seen in England; it lies below Lepus. Apart from Phakt, the only star of moderate brilliancy is β (Wezn), magnitude 3·2. There are no interesting objects in the group.

CÆLUM The Sculptor's Tools

This group (originally Cæla Sculptoris) was introduced by Lacaille in 1752. It lies near Columba, is always very low in Britain, and is entirely unremarkable. It is included in the map with Columba.

ARGO NAVIS The Ship Argo

Argo is one of Ptolemy's constellations, and represents the ship which carried Jason and his companions in their quest of the Golden Fleece. Unfortunately, nearly all of it is too far south to be seen in Britain. A few stars of that part of it known as Puppis (the Poop) do rise in our latitudes, of which the chief is the third-magnitude ρ. It is also possible to make out parts of two entirely unimportant constellations introduced by Lacaille in 1752, Pyxis Nautica (the Mariner's Compass) and Antlia (the Airpump), neither of which presents any features of interest. The visible part of Argo is included with the map of Canis Major (page 59).

CHIEF STAR

	Visual Magnitude	Spectrum	Absolute Magnitude	Distance, lt.-yrs.
ρ Turais	2·80	F6	0·3	105

SPRING STARS

The sky as seen on a spring evening lacks some of the glory of winter. Orion is sinking below the western horizon, and of his retinue only Capella, Procyon and the Twins are still reasonably high up. The Great Bear is practically overhead; Cassiopeia rather low in the north, with Vega easterly. The southern aspect is dominated by Leo, the Lion, while the brightest star on view is the glorious orange-coloured Arcturus, leader of Boötes the Herdsman. The Milky Way is not so conspicuous as in winter.

BOÖTES The Herdsman

An original constellation. Mythological accounts of it vary considerably. One story relates how Boötes was robbed by his brother of all his goods, and after wandering round the earth invented the plough, which was drawn by two oxen. His mother,

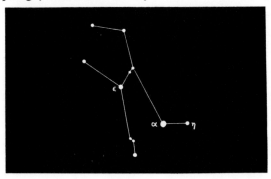

	Visual Magnitude	Spectrum	Absolute Magnitude	Distance, lt.-yrs.
α Arcturus	−0·06	K2	−0·3	36
ε Izar	2·37	K1	0·0	103
η Saak	2·69	G0	2·7	32
γ Seginus	3·05	A7	0·2	118

Callisto, was so impressed that she persuaded Jupiter to place Boötes in the sky.

Boötes is dominated by Arcturus, whose colour makes a splendid show in binoculars. The star is best located by following round the curve of the Great Bear's tail. Arcturus lies well north of the celestial equator, and is a prominent feature of the evening sky from spring through to autumn.

The constellation includes two notable doubles. δ, magnitude 3·5, has a companion of magnitude $7\frac{1}{2}$, just visible in good binoculars. ε (Izar) has a much closer companion of just below the sixth magnitude; the bright component is yellowish, the companion bluish. However, at least a 3-inch refractor is needed to show the companion.

CANCER The Crab

Cancer is a faint group, but it is in the Zodiac, and was included by Ptolemy. It represents a gigantic sea-crab which came to the help of the watersnake (Hydra) which was doing battle with Hercules. Not unnaturally, Hercules trod on the crab and killed it, whereupon Juno, queen of Olympus—who had her own reasons for being jealous of Hercules—placed it in the sky.

Despite the fact that its brightest stars (β and ι) are only of the fourth magnitude, Cancer is not hard to find, since it lies between the Lion and the Twins. It occupies the triangular area bounded

71

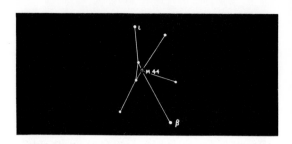

approximately by Pollux, Procyon and Regulus.
ι is a wide, easy double, but the most interesting
feature is the open cluster Præsepe, nicknamed the
Beehive, and numbered 44 in Messier's catalogue.
Præsepe is easily visible to the naked eye on a
moonless night, and binoculars give an excellent
view of it. Apart from the Pleiades it is the finest
loose cluster visible from Britain.

COMA BERENICES Berenice's Hair

A group introduced by the great Danish astronomer Tycho Brahe, who worked at his observatory
at Uraniborg, on the island of Hven, during the
latter part of the sixteenth century, and compiled
an excellent star-catalogue. A charming legend is
attached to it. When Ptolemy Euergetes, king of
Egypt, set off on a dangerous expedition against
the Assyrians, his wife, Queen Berenice, vowed that
if he returned safely she would cut off her beautiful
hair and place it in the temple of Venus. The king
duly returned; Berenice kept her promise, and
Jupiter placed the shining tresses among the stars.

Coma occupies the area bounded by Boötes,
Cor Caroli in Canes Venatici, and Denebola in
Leo. It contains no stars brighter than magnitude
$4\frac{1}{2}$, but there are numerous faint ones, and at a

72

casual glance the constellation resembles a large, scattered cluster. It is well worth sweeping with binoculars or a low power.

CORONA BOREALIS The Northern Crown

A small but conspicuous constellation, included in Ptolemy's list. It represents a crown given by Bacchus to Ariadne, daughter of King Minos of Crete.

CHIEF STAR

	Visual Magnitude	Spectrum	Absolute Magnitude	Distance. lt.-yrs.
α Alphekka	2·23	A0	0·4	76

Corona is one of the few groups which really bear some resemblance to the object after which they are named; the semicirclet of stars not far from Arcturus is very easy to identify. It contains a most peculiar star, T Coronæ, which is generally very faint, but which rose briefly to the second magnitude in 1866 and again to the third in 1946. It is classed as a 'nova-like variable', and is quite unpredictable, but generally it is invisible with small telescopes.

CORVUS The Crow

One of Ptolemy's groups. It is said that when the god Apollo fell in love with Coronis, the beautiful daughter of Phlegyas and mother of the great

73

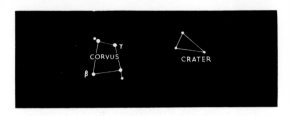

CORVUS

CRATER

CHIEF STARS

		Visual Magnitude	Spectrum	Absolute Magnitude	Distance, lt.-yrs.
γ	Minkar	2·59	B8	−3·1	450
β		2·66	G5	0·1	108
δ		2·97	B9·5	0·1	124
ε		3 04	K3	−0·2	140

doctor Æsculapius, he sent a crow to watch her. In the course of time the crow duly returned and made his report. Though his detective work had revealed activities which were hardly to Coronis' credit, Apollo rewarded the bird with a place in the sky.

These stars form a conspicuous little quadrilateral. Corvus is easy to find, rather low in the south during spring evenings; it lies not far from the first-magnitude Spica, in Virgo. It contains no objects of note.

CRATER The Cup

Crater is so inconspicuous that it is rather surprising to find that it is one of Ptolemy's original groups. It has been identified with the goblet of the wine-god Bacchus. It lies near Corvus, contains no star as bright as magnitude 3½, and is in no way important. It is included in the map with Corvus.

HYDRA The Watersnake or Sea-serpent

An original constellation. Mythologically it was a monster with a hundred heads, which infested the Lernæan marshes. For his second labour Hercules was ordered to kill it—which he duly did.

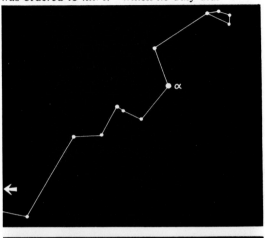

CHIEF STARS

		Visual Magnitude	Spectrum	Absolute Magnitude	Distance, lt.-yrs.
α	Alphard	1·98	K4	− 0·3	94
γ		2 98	G8	0·3	113

Hydra has the distinction of being the largest separate constellation in the sky. It is also one of the dullest. Its only bright star, Alphard, is aptly

known as 'the Solitary One'; it is easy to find
because a line drawn from Castor through Pollux
leads to it, and because of its strong orange hue.

The 'head' of Hydra lies near Cancer, and the
huge snake winds its way below Leo, Corvus and
Virgo. It is remarkably devoid of interesting
objects.

LEO The Lion

A Zodiacal constellation, and one of Ptolemy's
originals. Mythologically it is another of Hercules'
victims—the gigantic lion which hunted in the
Neméan forests.

This is a splendid group, and is the most conspicuous feature of the southern aspect during
spring evenings. The stars of the Great Bear act
as a direction-finder to it. Regulus is the brightest
star of the Sickle, which is shaped rather like a
question-mark as seen in a mirror; the rest of the

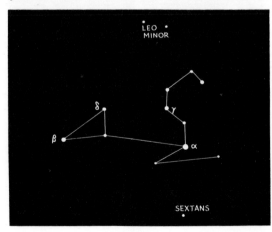

		Visual Magnitude	Spectrum	Absolute Magnitude	Distance, lt.-yrs.
α	Regulus	1·36	B7	−0·7	84
γ	Algeiba	1·99	K0	0·1	190
β	Denebola	2·14	A3	1·5	43
δ	Zozma	2·57	A4	0·6	82
ε	Asad Australis	2·99	G0	−2·1	340

bright stars in the group form a triangle. Denebola, ranked of the first magnitude by Ptolemy, is now below the second, but may be slightly variable. A small telescope will show that Algeiba is double; the companion is about 1½ magnitudes fainter than the primary. This is a binary system, with a period of 407 years.

LEO MINOR The Little Lion

A constellation formed by Hevelius in 1690. It seems to have no mythological associations, and neither does it contain any bright stars or objects of interest. It lies between the Great Bear and the Sickle. It is included in the map with Leo.

SEXTANS The Sextant

A faint group not far from Leo, formed by Hevelius in 1690—for no apparent reason, since it is extremely barren and obscure.

VIRGO The Virgin

A Zodiacal group, included in Ptolemy's list. According to the ancient Greek writer Hesiod, Virgo was identical with Astræa, the daughter of Jupiter and Themis, and the goddess of justice. During the Golden Age Astræa ruled the world, but when mankind altered its ways the goddess

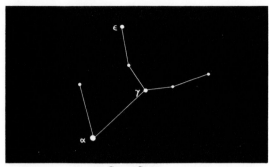

CHIEF STARS

	Visual Magnitude	Spectrum	Absolute Magnitude	Distance, lt.-yrs.
α Spica	0·91 (v)	B1	−3·3	220
γ Postvarta	2·76	F0	3·5	32
ε Vindemi-atrix	2·86	G9	0·6	90

was so disgusted that she returned to heaven. (Looking at the daily papers of A.D. 1983, one can easily understand how she must have felt!)

Spica, the leading star of the group, is best found by continuing the sweep from the Great Bear's tail through Arcturus. There are several other fairly bright stars in the constellation; note the 'bowl' shape which lies between Spica and Denebola in Leo. γ is a superb binary with almost equal components. At present it is easy to separate with a small telescope; but it is a system with a period of 180 years, and is gradually closing up as seen from Earth, so that during the next century it will become a difficult object to split.

The region enclosed by the 'bowl' and Denebola is very rich in galaxies, but unfortunately all these are faint objects visible only with large telescopes.

SUMMER STARS

Summer is generally regarded as the worst time for star-gazing. It is of course true that the period of darkness is relatively short, and moreover the brilliant winter constellations such as Orion are out of view. Yet the summer skies are certainly not devoid of interest.

On a late evening in July, for instance, the lovely blue star Vega is almost overhead, and takes up the position occupied in winter by Capella. Not far off are two more first-magnitude stars, Deneb in Cygnus and Altair in Aquila, while Arcturus may be seen to the west and Capella is low down near the northern horizon. In the south we have Antares in Scorpio, which is notable because of the strong red colour which has led to its being called 'the Rival of Mars'. The Great Bear occupies the western part of the sky, but Leo and Virgo have set. To the east, the Square of Pegasus is coming well into view. but will be better placed during autumn evenings. The Milky Way is prominent, extending in a glorious band from Capella through Cassiopeia, Cygnus and Aquila down to Scorpio.

AQUILA The Eagle

An original constellation. It is said to represent an eagle sent by Jupiter to carry off Ganymede, a shepherd-boy of Phrygia, whom the king of the gods desired to have as cup-bearer in place of his daughter Hebe—who awkwardly tripped and fell on a solemn occasion, and was consequently forced to resign her office.

79

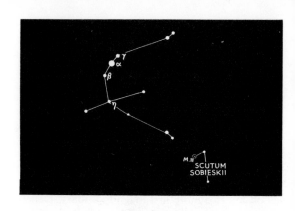

CHIEF STARS

		Visual Magnitude	Spectrum	Absolute Magnitude	Distance, lt.-yrs.
α	Altair	0·77	A7	2·2	16
γ	Tarazed	2·67	K3	−2·4	340
ζ	Dheneb	2·99	A0	0·8	90

Aquila is one of the most prominent of the summer groups. Its leading star, Altair, is white in colour, and is easily identified because it has a fainter star to either side of it: γ of slightly above the third magnitude, and β (Alshain) of the fourth. Altair is one of the closest of the first-magnitude stars.

The most interesting object in Aquila is η, a Cepheid variable with a period of 7·2 days and a magnitude range of from 3·7 to 4·5. Its fluctuations may be well followed with the naked eye.

CYGNUS The Swan

A splendid constellation, naturally included by Ptolemy. Several legends are associated with it.

According to one of these, Jupiter changed himself into a swan when he wanted to visit Leda, wife of the Spartan king Tyndarus, and—for reasons which are rather obvious—felt it desirable to remain strictly incognito. Later, in memory of the episode, he placed a swan in the sky.

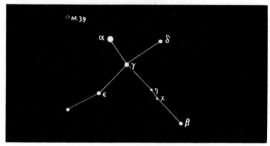

CHIEF STARS

		Visual Magnitude	Spectrum	Absolute Magnitude	Distance, lt.-yrs.
α	Deneb	1·26	A2	−7·1	1,600
γ	Sadr	2·22	F8	−4·6	750
ε	Gienah	2·46	K0	0·7	74
δ		2·87	B9·5	−1·7	270
β	Albireo	3·07	K3	−2·4	410

Cygnus is often, and more appropriately, termed the Northern Cross, and is so prominent that it may be identified at almost the first glance. One star of the cross—β or Albireo—is fainter than the rest, since it is below the third magnitude; but it makes up for this by being a superb double, with a golden–yellow primary and a greenish or bluish companion. Any small telescope will show it excellently, and it is probably the most beautiful object of its kind in the whole sky.

81

Deneb does not appear so conspicuous as Altair or Vega, since it lies at a great distance from us, but it is exceptionally luminous. It is a hot, white star.

Between γ, in the middle of the cross, and Albireo lies the star η Cygni, magnitude 4·0. Close to η is a long-period variable, χ, which changes between magnitudes 4 and 14, taking 409 days to pass from one maximum to the next. When at its brightest it may equal η; near minimum, telescopes of considerable size are needed to show it at all.

Since the Milky Way flows through Cygnus the whole constellation is rich, and sweeping with binoculars or a low power will reveal some magnificent star-fields. Note too the prominent open cluster M.39, which is well seen with a small telescope.

DELPHINUS The Dolphin

An original group. An attractive legend is associated with it. Arion, the famous lyric poet and musician, was sailing back from Corinth to Sicily when he was seized by the ship's crew, who coveted the prizes which he had won. Arion pleaded to be allowed to play one last tune on his cithara; this

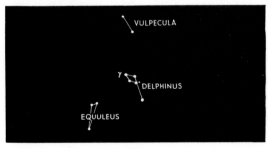

request was granted, and the music attracted a school of dolphins around the ship. Suddenly Arion flung himself into the sea, and one of the dolphins carried him safely to the port of Jænarius. It is this dolphin which we now see in the sky.

Delphinus is a small, compact group not far from Altair. It has no star as bright as the third magnitude, but it is nevertheless easy to recognize. A small telescope shows that γ is double, with a yellowish primary and a greenish companion of about magnitude $5\frac{1}{2}$.

EQUULEUS The Little Horse

Since Equuleus is so small and obscure, it is rather surprising to find it listed by Ptolemy. It represents a horse which Mercury, the messenger of the Gods, presented to Castor, one of the Heavenly Twins. It is included in the map with Delphinus.

HERCULES

An original constellation. Hercules is, of course, one of the most famous of the mythological heroes. He was the son of Jupiter and Alcmene, and was taught by the centaur Cheiron. Juno was jealous of him, and her cunning plans resulted in Hercules being subjected to the will of his half-brother Eurystheus, who ordered him to perform twelve difficult tasks—the 'Labours of Hercules'. All these were successfully completed, and when the great hero died he was made an immortal, becoming reconciled to Juno and marrying her daughter Hebe.

Oddly enough, Hercules is by no means prominent. It occupies a large area, more or less between Arcturus and Vega, but it contains no brilliant stars and has no obviously distinctive

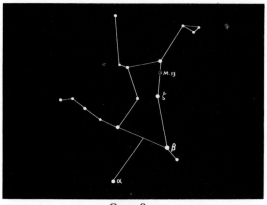

CHIEF STARS

	Visual Magnitude	Spectrum	Absolute Magnitude	Distance, lt.-yrs.
β Korne-phoros	2·78	G8	0·3	103
ζ Rutilicus	2·81	G0	3·1	30

shape. It does, however, contain two very interesting objects. One is the red M-type variable α (Rasalgethi), which fluctuates irregularly between magnitudes 3 and 4. The nearby star κ Ophiuchi, magnitude 3·4, is a good comparison, and though the changes are slow they may be followed with the naked eye. Rasalgethi is of immense diameter, and may indeed be amongst the largest stars known to us.

Even more fascinating is Messier 13, which may just be seen with the naked eye on a clear night, and is obvious in binoculars. This is a globular cluster, and contains something like 100,000 stars. It is remote, and its distance has been estimated as 34,000 light-years.

About 100 globular clusters are known. Most of them are in the southern part of the sky, and M.13 is indeed the only really conspicuous globular visible from Britain. They form a kind of outer surround to our Galaxy, and their lop-sided distribution enabled the American astronomer Harlow Shapley, over 60 years ago, to show that the Sun lies well away from the galactic centre.

M.13 is not spectacular when viewed with binoculars, but a moderate telescope will resolve its outer parts into separate stars, and with large apertures it is a fine sight.

LIBRA The Scales or Balance

A Zodiacal constellation listed by Ptolemy, but apparently known previously as Chelæ Scorpionis (the Scorpion's Claws). Greek legends state that it commemorates Mochis, the inventor of weights and measures, but no well-defined mythological stories seem to be associated with it.

Libra is decidedly obscure. It lies between Spica and Antares, and is best seen during evenings in early summer. α has a fifth-magnitude companion visible to the naked eye. β is said to be the only naked-eye star with a perceptibly greenish hue, though to my eyes it always seems white.

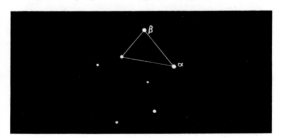

	Visual Magnitude	Spectrum	Absolute Magnitude	Distance, lt.-yrs.
β Zubenel-chemale	2·61	B8	−0·6	140
α Zubenel-genubi	2·76	A3	1·2	66

LYRA The Lyre

An original constellation. Despite its small size it is of exceptional interest. It represents the lyre presented by Apollo to Orpheus, son of Eagrus and Calliope. With this lyre Orpheus charmed not only wild beasts but also the stones and trees, and even chained the rivers in their courses. After Orpheus' death his lyre was placed in the sky.

CHIEF STAR

	Visual Magnitude	Spectrum	Absolute Magnitude	Distance, lt.-yrs.
α Vega	0·04	A0	0·5	26

Vega is the fifth brightest star in the sky; of those visible in Britain it is surpassed only by Sirius and Arcturus. It is of a lovely blue colour, and is almost overhead during summer evenings, so that it cannot be mistaken.

Close to Vega is the famous multiple ε. With the naked eye the two main components may be separated, while a 3-inch refractor shows that each component is again double. Its neighbour ζ is also a binary separable with a small telescope.

Sheliak or β Lyræ is an eclipsing variable changing between magnitude 3·4 and 4·1; its period is almost 13 days. Unlike Algol in Perseus it is always varying, because both stars are of high luminosity; whereas in Algol the second component is dim, so causing a marked drop in light when it hides the brilliant component. The components cannot be seen separately; they are so close together that they almost touch, and each is drawn out into an elliptical or egg-like shape, though of course no telescopes will show them as such. The changes in brightness are easy to detect by using the nearby star γ Lyræ, magnitude 3·0, as a comparison.

Midway between β and γ is the Ring Nebula, Messier 57. It is not bright, and at least a 3-inch telescope is needed to show it unmistakably. With larger apertures it is seen to take the form of a luminous ring with a very faint star in the centre. It is the brightest of a class of objects known as planetary nebulæ, but the name is misleading, since it is neither a planet nor a nebula. It is in fact a faint, very hot star which is surrounded by an immensely distended 'atmosphere' or shell of tenuous gas.

OPHIUCHUS The Serpent-bearer

Ophiuchus, listed by Ptolemy, is not counted as a Zodiacal constellation, though part of it does extend into the Zodiac between Scorpio and Sagittarius. Mythologically Ophiuchus—sometimes called Serpentarius—is identified with Æsculapius, son of Apollo and Coronis, who became so skilled in medicine that he not only healed the sick but also restored the dead to life.

Chief Stars

		Visual Magnitude	Spectrum	Absolute Magnitude	Distance, lt.-yrs.
α	Rasalhague	2·09	A5	0·8	58
η	Sabik	2·46	A2·5	1·4	69
ζ	Han	2·57	O9·5	−4·3	520
δ	Yed Prior	2·72	M1	−0·5	140
β	Cheleb	2·77	K2	−0·1	124

This alarmed Pluto, ruler of the Underworld, who feared that his kingdom would become de-populated. Rather reluctantly Jupiter killed Æsculapius with a thunderbolt, but afterwards relented sufficiently to place him in the sky.

Ophiuchus is a large constellation, lying between Vega and Antares. Rasalhague lies near Rasal-gethi in Hercules. Despite its considerable area

Ophiuchus is dull and barren, and there are few objects of interest in it.

SAGITTA The Arrow

A little constellation listed by Ptolemy, and often identified with Cupid's bow. It is said, too, to be the arrow with which Apollo exterminated the one-eyed Cyclops, and there are various other legends about it. It lies between Aquila and Cygnus, not far from Altair. It is compact and not hard to identify, but contains no stars as bright as the third magnitude, and no objects of note.

SAGITTARIUS The Archer

An original group. It has been identified with the centaur Chiron, tutor of Jason and Hercules, but according to another story Chiron merely invented the constellation to act as a guide to the Argonauts in their quest of the Golden Fleece. Sagittarius is the most southerly of the Zodiacal

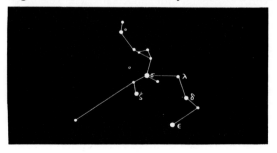

	Visual Magnitude	Spectrum	Absolute Magnitude	Distance, lt.-yrs.
ε Kaus Australis	1·81	B9	−1·1	124
σ Nunki	2·12	B2	−2·7	300
ζ Ascella	2·61	A2	0·1	140
δ Kaus Meridionalis	2·71	K2	0·7	84
λ Kaus Borealis	2·80	K2	1·1	71
γ Alnasr	2·97	K0	0·1	124
π Albaldah	2·89	F2	−0·7	250

constellations. It is always very low in Britain, and part of it remains permanently below our horizon.

The group is not striking. The best method of finding it is to locate Antares in Scorpio and then look to the east, about the same distance above the horizon. The brightish stars which mark Sagittarius should then be found without a great deal of trouble.

There are some interesting objects in the area, but their low altitude above the British horizon prevents their being seen well. However, the area is worth sweeping with binoculars. The celebrated 'star clouds' indicate the direction of the centre of the Galaxy.

SCORPIO The Scorpion

The eighth constellation of the Zodiac, and an original group. Mythologically it is the scorpion which rose out of the ground, at Juno's command, to attack the great hunter Orion. Another legend associates it with the story of Phæthon, the rash boy who obtained permission to drive the sun-chariot for one day. It is related that the horses,

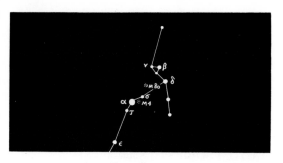

CHIEF STARS

		Visual *Magnitude*	Spectrum	*Absolute* *Magnitude*	*Distance,* *lt.-yrs.*
α	Antares	0·92 (v)	M1	−5·1	66
ε	Wei	2·28	K2	0·7	66
δ	Dschubba	2·34	B0	−4·0	590
β	Graffias	2·63	B0·5	−3·7	650
τ		2·85	B0	−4·0	750
σ		2·86	B1	−4·4	570
π		2·92	B1	−3·3	570
μ		2·99	B1·5	−3·0	520

already restive and unruly in their journey across the sky, were startled when they came across the celestial scorpion with its sting upraised, and the inexperienced driver lost control of them completely, so that Jupiter was forced to strike him with a thunderbolt.

Scorpio is a superb constellation, but is too far south to be well seen in Britain, and several of its bright stars—notably λ (Shaula), magnitude 1·7 —remain below our horizon. Antares may, however, be seen during summer evenings, recognizable because of its brightness and because of its strong red colour. It is exceptionally large, and its diameter has been estimated as 350,000,000 miles.

Like Altair it has a fainter star to either side of it—
in this case τ and σ, each of about the third magnitude. Moderate telescopes show that Antares has a
greenish companion of magnitude 7.

ν is a wide, easy double, separable in good
binoculars, and there are also two bright clusters
in the constellation: M.80, a globular some 65,000
light-years from us, conspicuous in a small
telescope; and M.4, a fainter loose cluster between
Antares and σ.

SCUTUM SOBIESKII Sobieski's Shield

A small group, introduced by Hevelius in 1690
in honour of John Sobieski, king of Poland, who
defeated the Turks under the walls of Vienna. The
constellation represent's Sobieski's coat of arms.

Scutum lies near Aquila, and is included in the
map of the constellation (page 80). It contains no
star as bright as the fourth magnitude, but there is
one interesting loose cluster—M.11, known commonly as the 'Wild Duck' cluster. It is dimly
visible with the naked eye, and is a fine sight in a
telescope.

SERPENS The Serpent

An original constellation, but a most confusing
one. It is the serpent with which Ophiuchus is
struggling, and has evidently had the worst of the
battle, since it has been pulled in half and now
consists of two quite unconnected parts—Caput
(the Head) and Cauda (the Tail)—with Ophiuchus
in between. It has therefore been included in the
map of Ophiuchus (page 88), though from its size
it certainly merits a chart to itself.

The only notable object is the fine double θ
(Alya), which lies in Cauda, not far from δ Aquilæ;
Aquila, indeed, represents the best pointer to it.

	Visual Magnitude	Spectrum	Absolute Magnitude	Distance, lt.-yrs.
α Unukalhai	2·65	K2	1·0	71

Both components of θ are of about magnitude $4\frac{1}{2}$, and they are widely separated, so that any small telescope will show them well.

VULPECULA The Fox

Introduced by Hevelius in 1690. Originally it was known as Vulpecula et Anseris, the Fox and Goose, but nowadays the Goose is generally forgotten. The group of extremely obscure, and contains no objects of interest to the modestly equipped observer. It lies between Delphinus and Cygnus, and is included in the map of Delphinus (page 82).

AUTUMN STARS

Autumn is a relatively barren time of the year so
far as the evening sky is concerned. The bright
summer groups such as Cygnus and Aquila are
still to be seen, but are becoming low in the west;
Antares has disappeared, and so too has Arcturus.
Orion is still invisible, but Taurus is well above
the horizon; the lovely star-glow of the Pleiades
has become prominent in the east, and Capella is
conspicuous once more. The Plough is at its
lowest, in the north, with Cassiopeia almost overhead. The southern aspect is dominated by the
Square of Pegasus. Much of the south part of
the sky is however occupied by large, faint constellations such as Cetus and Aquarius.

PEGASUS The Flying Horse

It is best to begin our autumn survey with Pegasus,
since this is the leading constellation of the season.
It is an original constellation, and honours the
winged steed upon which Perseus rode on his way
back from his successful Gorgon-slaying expedition. It was then that he saw Andromeda chained
to the rock, and swooped down to rescue her.
Later Pegasus was given to another hero, Bellero-
phon, to help in conquering the Chimæra, a
hideous, three-headed, fire-breathing monster.
Having dealt faithfully with the Chimæra, Bellero-
phon decided to ride up to Olympus; but Jupiter,
angered at his coolness, sent a gad-fly to sting
Pegasus and make him dismount his rider. Pegasus
himself continued the upward journey, and was
duly placed among the stars.

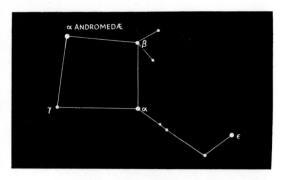

	Visual Magnitude	Spectrum	Absolute Magnitude	Distance, lt.-yrs.
ε Enif	2·31	K2	−4·6	780
β Scheat	var.	M2	−1·5	210
α Markab	2·50	B9·5	−0·1	109
γ Algenib	2·84	B2	−3·4	570
η Matar	2·95	G8	−2·2	360

Three of the chief stars of Pegasus (α, β and γ), together with Alpheratz or α Andromedæ, make up the celebrated Square. Actually the Square is not so prominent as might be thought from the maps, but it is easy enough to find, as it is high in the south during autumn evenings. Two of the stars in the W of Cassiopeia point to it. The most interesting object is the red giant β, which varies between magnitudes 2¼ and 2¾ in a period which is roughly 35 days, but subject to marked irregularities. Its diameter is nearly 150,000,000 miles, but its mass is only 9 times that of the Sun, so that—like all red giants—it is comparatively rarefied. Its changes may be followed by using α and γ as comparison stars.

95

ANDROMEDA

Andromeda, contained in Ptolemy's list, is one of the chief characters in the Perseus legend.

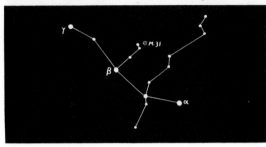

		Visual Magnitude	Spectrum	Absolute Magnitude	Distance, lt.-yrs.
β	Mirach	2·02	M0	0·2	76
α	Alpheratz	2·06	B9p	−0·1	90
γ	Almaak	2·14	K3	−2·4	260

Alpheratz is included in the Square of Pegasus, and was formerly known as δ Pegasi; the reason for its transfer is not clear. Andromeda consists chiefly of a line of fairly bright stars extending from the Square towards Perseus.

γ is a fine double. The primary is yellowish, and a moderate telescope will suffice to show the fifth-magnitude bluish companion. The most interesting object in the constellation is, however, Messier 31—the Great Spiral, one of the nearest of the outer galaxies. On a clear night it is just visible to the unaided eye, not far from the 4½-magnitude star ν, and binoculars or a small telescope will show it clearly. However, it must

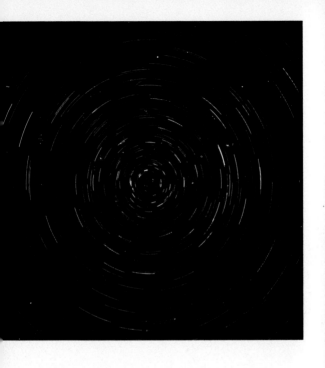

Plate 1   Star trails. The camera is pointed at the pole, and a time exposure made

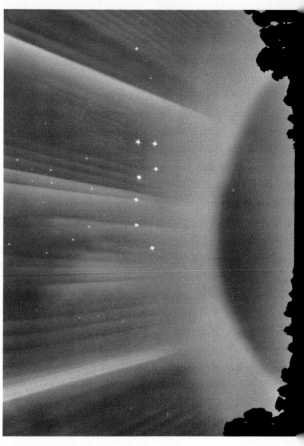

Plate 2 A fine display of the Aurora Borealis as seen from southern England in November 1961. Note the constellation of Ursa Major (the Plough) near the centre

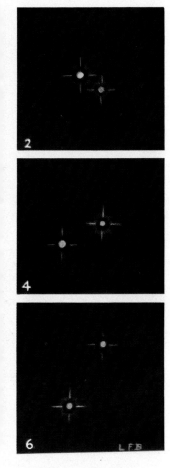

stars as seen through a 3-inch
[...]lis. 2. 70 Ophiuchi. 3. Alpha
[...]medae. 5. Gamma Arietis. 6.
[...]r rays are not objective, but
[...]mings of both eye and telescope

Plate 4 Spiral galaxy M.101 in Ursa Major

Plate 5 *Upper:* The Pleiades
Lower: The Sword-handle in Perseus: a double cluster

Plate 6 *Upper:* The planet Mercury
Lower: The planet Venus

Plate 7 An impression of the conspicuous comet Arend-Roland in the northern sky during April 1957. The cometary head lies close to the star Alpha Persei

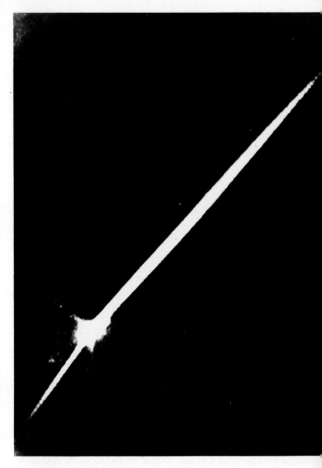

Plate 8 Exploding Andromedid meteor

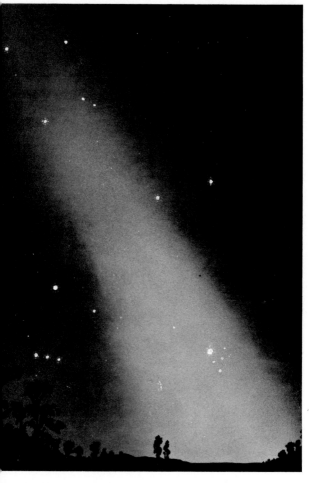

Plate 9 The Zodiacal Light as seen from northern India in the spring of 1943

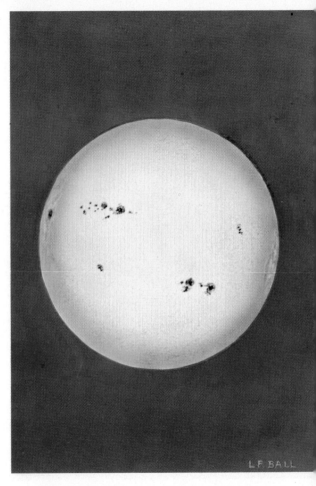

L.F. BALL

Plate 10 The Sun as seen through a small telescope by projection, showing sunspot groups and faculæ at the limb

Plate 11 The planet Jupiter
Upper: 1957 March 20, showing the Great Red Spot
Lower: 1935 May 22, with great activity on the equatorial belts

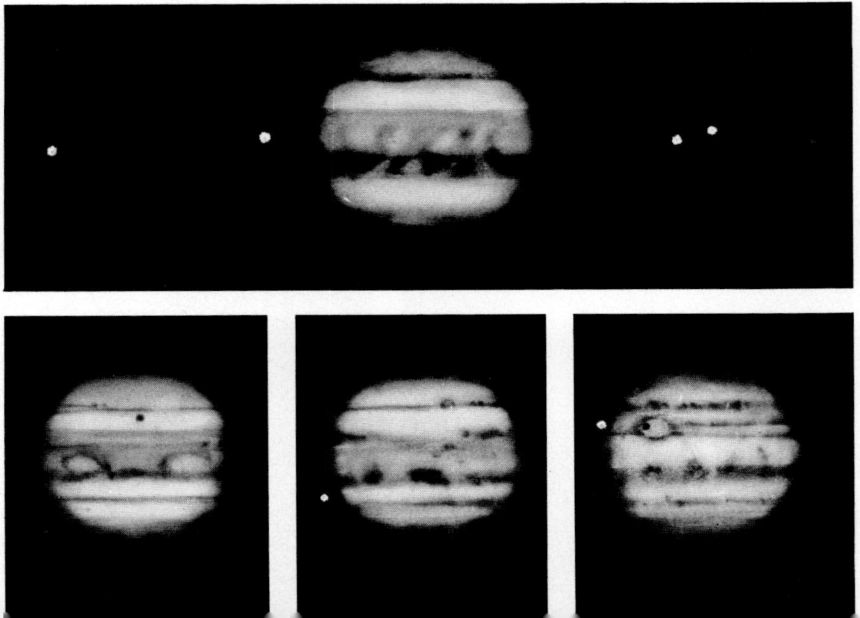

Plate 12
Jupiter's satellites.
Upper: General
view of the planet
and its four major
moons
Lower: Left,
transit of
Ganymede.
Centre, Io about to
be occulted. Right,
Io and its shadow

Plate 13
Supernova in galaxy M.101 in Ursa Major
Left: Without nova, 1950 June 9
Right: With nova, 1951 February 7

FEB. 7, 1951

JUNE 9, 1950

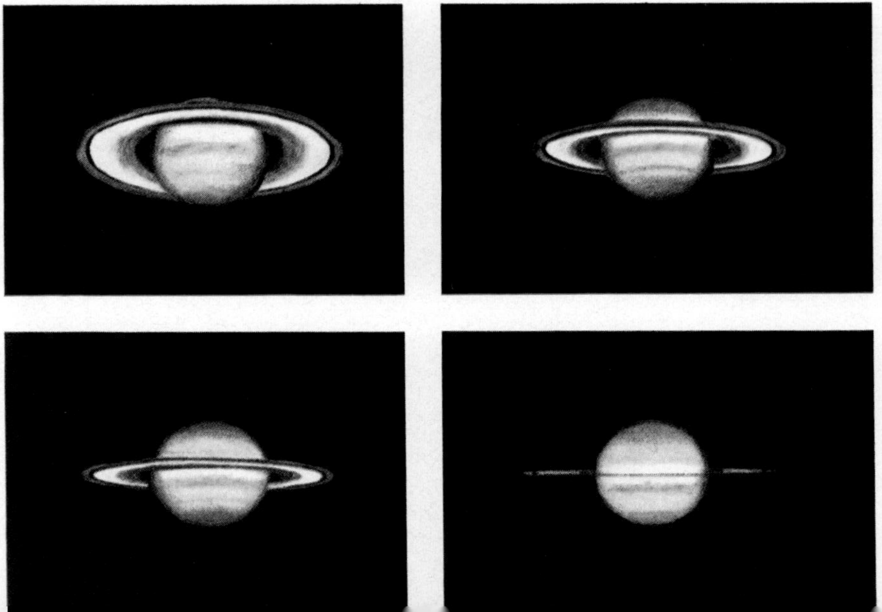

Plate 14
Varying
aspects of
Saturn's rings.
Upper: Left,
1955; right,
1953
Lower: Left,
1952; right,
1950

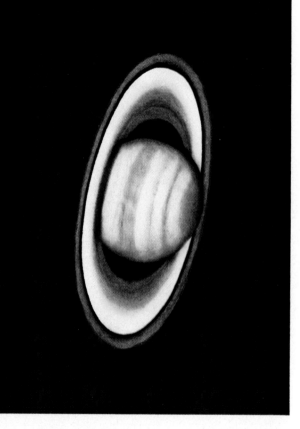

Plate 15 The planet Saturn with its ring system almost fully open as presented to the Earth in June 1958

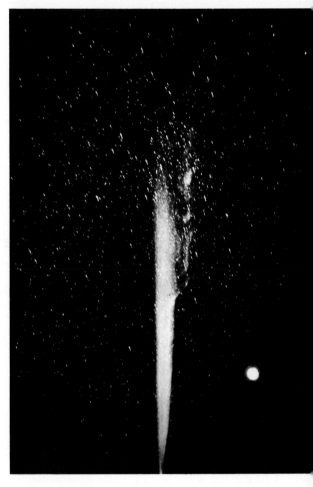

Plate 16   Halley's Comet and Venus

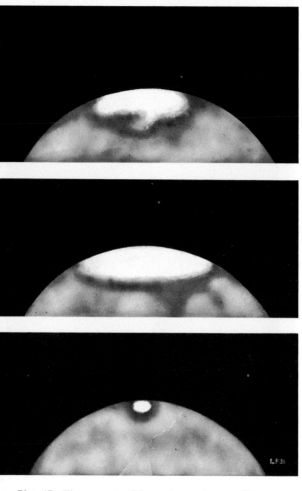

Plate 17   Three aspects of the southern polar cap of Mars
showing variation in size during the apparition of 1956

Plate 18
Mars, 1971;
G. Roberts,
26-inch
refractor,
Republic
Observatory,
Johannesburg

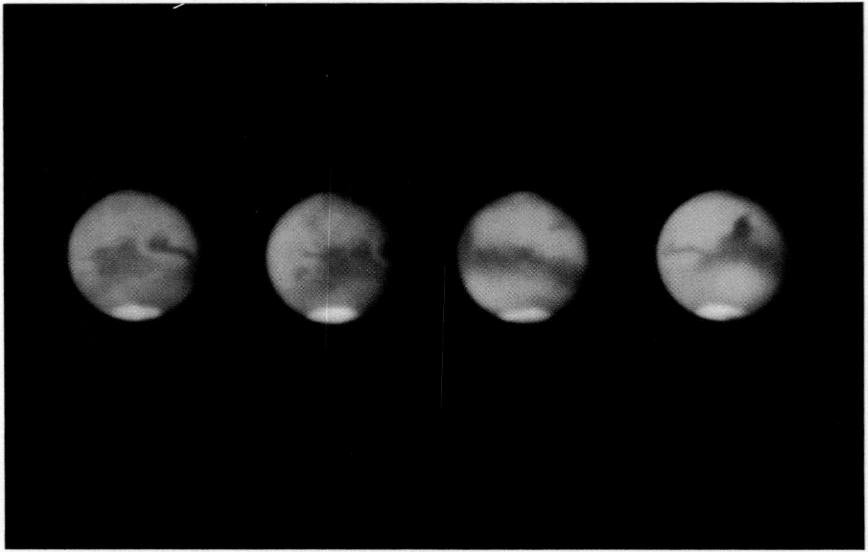

Plate 19
Mars, from
Viking 1:
August 1976.
The
spacecraft's
scoop is seen
in its parked
position.
Large blocks,
3 to 6 feet
across, can be
seen on the
horizon just
over 100
yards from
the spacecraft

Plate 20 Venus, photographed from Mariner 10 in February 1974. The cloud-patterns are clearly shown

Plate 21 Venus from Venera 9. This was the first picture ever to be received from the surface of Venus; the Russian description of the scene as ' a stony desert' is quite appropriate! Venera 9 landed on Venus on 22 October 1975. Most of the rocks shown are from 2 to 4 feet across. Only one picture was received before the probe was put permanently out of action by the intensely hostile conditions. Venera 10, which landed on the planet on 27 October 1975, also sent back one picture, showing a landscape of the same general type. Veneras 13 and 14 sent back even better pictures in 1981.

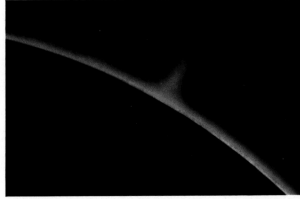

Plate 22 *Upper:* Projecting the Sun; Ronald Makins holds a white card behind the eye-piece of a 3-inch refractor
Lower: Solar Prominence, photographed by W. M. Baxter (4-inch refractor, Lyot filter)

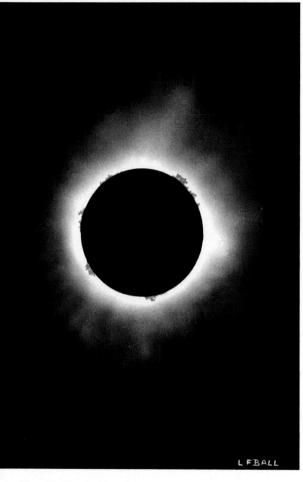

L F BALL

Plate 23 Total solar eclipse

Plate 24 The eastern half of the Moon

Plate 25   The western half of the Moon

Plate 26 Full moon

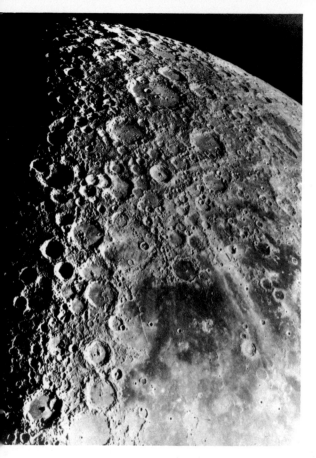

Plate 27 The south-east part of the Moon. Ptolemaeus is the large crater at the bottom left; above it are Alphonsus and Arzachel. The Straight Wall is to the right of Arzachel. Clavius, with its inner line of craters, is near the top

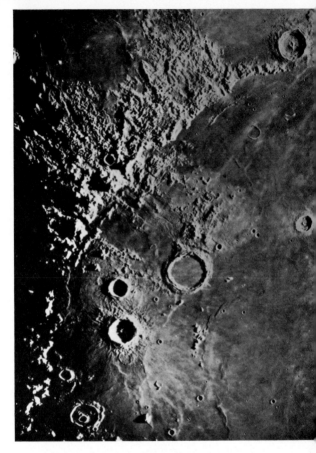

Plate 28   The Archimedes region of the Moon. The lunar Apennines are shown. Archimedes lies near the centre of the plate, with Eratosthenes near the top right

Plate 29 The Sinus Iridum (Bay of Rainbows) on the Moon, photographed by Patrick Moore with his 15-inch reflector on 18th August 1976

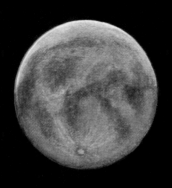

Plate 30 An eclipse of the Moon. The red hues are produced by refraction of the Sun's rays through the Earth's atmosphere
Upper: Mid eclipse
Lower: The Earth's shadow retreating

Plate 31
The annular eclipse of 1976, April 29; a series of
photographs taken by Patrick Moore from the Greek
Island of Thera

Plate 32 On the Moon — the scene from Apollo 15, with the lunar module and, to the right, the LRV or Lunar Rover

be admitted that with average amateur equipment it is decidedly disappointing, and looks only like a faint 'fuzz' of light. Large instruments are needed to show it in its true guise.

ARIES The Ram

This is still regarded as the first constellation of the Zodiac—though owing to the precession of the equinoxes the so-called First Point of Aries, or Vernal Equinox, has now shifted into the neighbouring group of Pisces. Aries is, of course, an original group.

It is said that Athamas, king of Thebes, had two children, Phryxus and Helle. Both children were badly treated by their stepmother, Ino, and the matter came to the attention of Mercury, messenger of the gods. Learning that Ino proposed to kill Phryxus, Mercury sent a ram with a golden fleece to rescue them. The ram could fly, and bore the two children across the water. Unfortunately Helle slipped and fell, and was drowned in the sea which is now known as the Hellespont or 'Sea of Helle'. Phryxus managed to hold on until he reached safety. After the ram's death the golden fleece was hung in a sacred grove, and was later removed by Jason and his Argonauts, while the ram itself was honoured by being given a place in the sky.

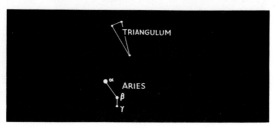

		Visual		*Absolute*	*Distance,*
		Magnitude	*Spectrum*	*Magnitude*	*lt.-yrs.*
α	Hamal	2·00	K2	0·2	76
β	Sheratan	2·68	A5	1·7	52

Aries lies below Andromeda, and is not very conspicuous. The most notable object is γ (Mesartim), a splendid double well seen with a small telescope. The components are each of about magnitude 4¼.

AQUARIUS The Water-bearer

An original group, and the eleventh constellation of the Zodiac. No well-defined legends are attached to it. According to some authorities it represents Ganymede, who succeeded Hebe as cup-bearer of the gods; the Egyptians imagined that its setting caused the Nile to rise.

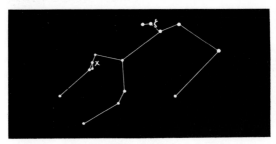

CHIEF STAR

		Visual		*Absolute*	*Distance,*
		Magnitude	*Spectrum*	*Magnitude*	*lt.-yrs.*
β	Sadalsuud	2·86	G0	−4·6	780

Aquarius is obscure, though it covers a large area. The best way to find it is to use two of the stars in the Square of Pegasus, β and α, as pointers. The group of faint yellowish stars round χ Aquarii is worth looking at, and ζ is double, but there are few objects in the constellation which will interest the beginner. Aquarius contains only one star as bright as the third magnitude.

CAPRICORNUS The Sea-goat

Another large, dull Zodiacal constellation. It has been identified with the demigod Pan, who was associated with shepherd and pastoral life in general, while according to one Greek legend it was 'the Gate of the Gods' through which the souls of men ascend to heaven. Capricornus contains no bright stars, and is not too easy to identify; it lies roughly between Altair and the southern first-magnitude star Fomalhaut.

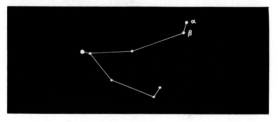

CHIEF STAR

	Visual Magnitude	Spectrum	Absolute Magnitude	Distance, lt.-yrs.
δ	2·92	A6	2·0	50

There are two objects of some interest, α (Giedi) is a naked-eye double; its components are of magnitudes 3·7 and 4·3. Its neighbour β (Dabih) is also double; magnitudes 3·3 and 6, separable with binoculars.

99

CETUS The Whale or Sea-monster

An original constellation, representing the monster of the Perseus legend.

Cetus is a large group, but is rather faint. It lies below the brighter constellations of the Andromeda region; its brightest star, β, is rather isolated. α, in the 'head', is decidedly reddish.

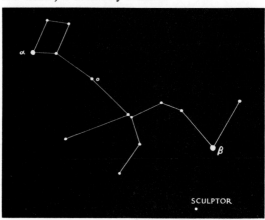

CHIEF STARS

	Visual Magnitude	Spectrum	Absolute Magnitude	Distance, lt.-yrs.
β Diphda	2·02	K1	0·8	57
α Menkar	2·54	M2	−0·5	130
o Mira	var.	M6	−0·5	103

The most interesting object is Mira (o Ceti), the most famous of all variable stars. At minimum it is only of the ninth magnitude, so that a moderate telescope is needed to show it, but at maximum it is easily visible to the naked eye. It has been known to surpass Polaris, but generally its greatest yearly magnitude does not exceed 3. The period is about 331 days, but is subject to fluctuations. Generally Mira is visible without a telescope for only about 18 weeks of its 47-week period. It is a red giant of vast diameter, and lies at the distance of 250 light-years. Forecasts for it are given, as accurately as possible, in various annual handbooks, and it is always worth while searching for Mira for a few weeks to either side of the predicted date of maximum. Sometimes, of course, maximum occurs when the star is not above the horizon except during daylight.

PISCES The Fishes

An original group; the last constellation of the Zodiac, though it now contains the Vernal Equinox. There is a mythological legend according to which Venus and Cupid once escaped from the giant Typhon by throwing themselves into the River Euphrates and changing themselves into fishes. Subsequently Minerva commemorated their escape by placing two fishes in the sky. This legend has also been associated with the Southern Fish, Piscis Australis.

Pisces consists mainly of a line of faint stars running below the Square of Pegasus. It is both obscure and uninteresting.

PISCIS AUSTRALIS The Southern Fish

Also called Piscis Austrinus. An original constellation, though Fomalhaut is its only bright star.

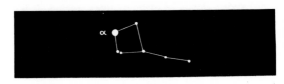

CHIEF STAR

	Visual Magnitude	Spectrum	Absolute Magnitude	Distance, lt.-yrs.
α Fomalhaut	1·19	A3	2·0	22·6

Fomalhaut is the most southerly of the first-magnitude stars visible in Britain, and may become quite conspicuous during autumn evenings. It is best found by using β and α Pegasi, in the Square, as pointers. It is white in colour, and rather isolated. Piscis Australis contains nothing else of interest.

SCULPTOR The Sculptor

First drawn by Lacaille in 1752, and once known as Apparatus Sculptoris (the Sculptor's Apparatus). It lies near β Ceti, and is extremely barren, with no star as bright as the fourth magnitude. It is included in the map with Cetus (page 100).

TRIANGULUM The Triangle

A small constellation near Andromeda, included in Ptolemy's list. The name is apt, since its three chief stars—all below the third magnitude—do indeed form a triangle. It contains nothing of interest to the casual observer. It is included in the map with Aries (page 97).

This, then, is a very brief outline of the constellations to be seen in British skies. Yet there are some spectacular groups too far south to rise above our horizon, so let us turn next to the stars of the southern heavens.

SOUTHERN STARS

The southern stars are just as interesting as the constellations of the north—indeed, perhaps more so; there is nothing in the northern sky to match the Clouds of Magellan, the brilliant twins Alpha and Beta Centauri and of course that symbol of the Antipodes, the Southern Cross. On the other hand there is no conspicuous south polar star, and the dim Sigma Octantis, magnitude 5·5, is no substitute for our Polaris.

Let us suppose that we are looking at the sky at nine o'clock in the evening from the latitude of Sydney, or the northern part of New Zealand. What will be seen—and shall we be able to find any of the familiar northern groups?

The south celestial pole lies in a rather barren area, surrounded by small, faint constellations with strangely modern names. Indeed, few of the southernmost groups have any mythological legends attached, simply because the old star-cataloguers knew nothing about them. Various astronomers have added new groups from time to time, not always with sound judgment. Some suggested constellations, with barbarous names such as Taurus Poniatowskii, Sceptrum Brandenburgicum, Officiana Typographica and Globus Aërostaticus, have been mercifully dropped from the maps. On the other hand the huge, rather unwieldy constellation of Argo Navis has been divided up into Carina (the Keel), Vela (the Sails) and Puppis (the Poop).

If we start our observations at 9 p.m. on 1st January, we shall see Orion high up, slightly to the east of north; Britons will have to do some mental readjustment at seeing Rigel higher than Betelgeux, with Sirius higher than either! The glorious supergiant Canopus is east of the zenith; Capella skirts the northern horizon; Aldebaran and the Pleiades are on view, and Castor and Pollux rising, though neither can rise very high in these latitudes. Over to the west lie Cetus and Eridanus. Eridanus can, of course, be seen in its entirety, with its prominent leading star Achernar.

APUS The Bird of Paradise

In his star-maps of 1603, Bayer added two small southern constellations: Apus (the Bee) and Avis Indica (the Bird of Paradise), only one of which has survived; the two are often confused. The modern Apus is an insignificant group close to the South Pole; its only moderately bright star is α, magnitude 3·8. θ is an M-type irregular variable, with a magnitude range of from 5·0 to 6·6. Apus is included in the map with Octans (page 109).

CHAMÆLEON The Chameleon

Another of Bayer's constellations, in the same area as Apus and even less conspicuous—it contains no star as bright as the fourth magnitude. It is included in the map with Octans (page 109).

DORADO The Swordfish

Dorado, also introduced by Bayer, has only two stars above the fourth magnitude—the more conspicuous, α, is ranked 3·5, and lies not very far

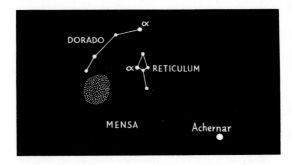

from the brilliant Achernar in Eridanus. It is
notable because it contains part of the Nubecula
Major or Great Magellanic Cloud, though part of
the Cloud extends into Mensa. The two Nubeculæ
are the closest of the external systems; the Great
Cloud looks almost like a detached portion of the
Milky Way, and is visible to the naked eye even in
strong moonlight. With binoculars or any telescope
it is a superb sight.

ERIDANUS The River

Eridanus is a long, sprawling constellation. Part
of it is of course visible from Britain—β (Kursa)—
lies near Rigel—but the two most interesting stars,
α and θ, are too far south to be seen from anywhere
in Europe.

Achernar is just circumpolar in the southern
part of Australia and New Zealand, and its
brilliance makes it unmistakable at once. Acamar
is of interest, since it was ranked of the first
magnitude by the old astronomers, and may
possibly have faded during the last 2,000 years.
Acamar is also a splendid binary, separable with a
modest telescope.

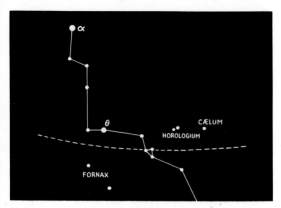

	Visual Magnitude	Spectrum	Absolute Magnitude	Distance, lt.-yrs.
α Achernar	0·51	B5	−2·3	118
θ Acamar	2·92	A3	1·7	65

HOROLOGIUM The Clock

A constellation added by Lacaille in 1752—though without obvious reason, since there is nothing to justify its independence. Its brightest star, α, is only of magnitude 3·8. Horologium is included in the map with Eridanus.

HYDRUS The Little Snake

One of Bayer's constellations, added in 1603. No mythological legends attach to it, and the name is perhaps unfortunate, since there is an obvious danger of confusion with Hydra. The official abbreviations used by the International Astronomical Unions are 'Hya' for Hydra and 'Hyi' for Hydrus.

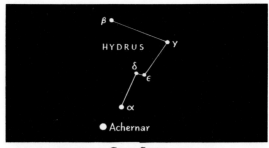

	Visual Magnitude	Spectrum	Absolute Magnitude	Distance. lt.-yrs.
β	2·78	G1	3·7	21
α	2·84	F0	2·9	31

Hydrus is circumpolar in south Australia and New Zealand; it lies roughly between Achernar and the South Pole, and α Hydri is close to Achernar. There are no objects here of interest to the amateur.

MENSA The Table

Introduced by Lacaille in 1752, under the name of Mons Mensæ (the Table Mountain), but now shortened to Mensa. It has no star as bright as the fifth magnitude, but part of the Nubecula Major extends into it. Mensa is included in the map with Dorado (page 106).

MUSCA AUSTRALIS The Southern Fly

A constellation between Chamæleon and the Southern Cross.

α is very slightly variable. β is double; the components are not very unequal in magnitude (3·5, 4·1) but are rather close together.

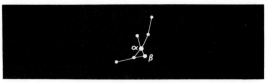

	Visual Magnitude	Spectrum	Absolute Magnitude	Distance, lt.-yrs.
α	2·70	B3	−2·9	430
β	3·06	B3	−2·1	470

OCTANS The Octant

Octans contains the south celestial pole, but is a remarkably barren constellation; its brightest star, *ν*, is only of magnitude 3·7, while the official pole star, *σ*, is well below the fifth magnitude. The longer arm of the Southern Cross is a rough guide to the pole. Octans is one of Lacaille's groups.

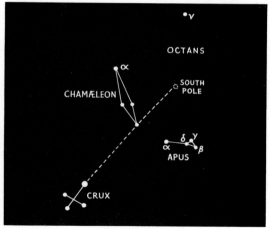

RETICULUM The Net

Lacaille, who introduced the group in 1752, gave it the name of Reticulum Rhomboidalis (the Rhomboidal Net), but the present shortened form is certainly an improvement! Reticulum is a compact little constellation (circumpolar in south Australia) between Dorado and Achernar, not far from the Nubercula Major. Its chief star, α, is of magnitude 3·4.

APRIL STARS

On April evenings Orion is approaching the western horizon, but Sirius is still prominent; Castor and Pollux are in the north-west, low down, and Regulus with Leo in the north. Spica appears toward the east, and Arcturus too is above the horizon. Keen eyes can make out a few stars of Ursa Major low in the north, but the Plough itself never rises from New Zealand or south Australia.

Centaurus and the Southern Cross are high in the eastern part of the sky, not far from the zenith, and could be classed as 'April stars' or 'July stars'; the Cross is, of course, circumpolar. The great Ship, Argo, is now at its best, and all of the long, dreary constellation of Hydra can be seen, stretching across the northern sky from Cancer in the west over to Libra in the east.

ARGO NAVIS The Ship Argo

Argo, in mythology the ship which carried Jason and his comrades on their successful quest of the Golden Fleece, was one of Ptolemy's original groups. Part of it, including the third-magnitude star, ρ, rises in Britain, but the most glorious part is too far south to be seen from anywhere in Europe.

110

Argo is so large that it is frankly unwieldy, and it has now been divided into separate parts: Carina (the Keel), Puppis (the Poop) and Vela (the Sails). Canopus, which is second in apparent brightness only to Sirius, is included in Carina, which is the richest part of the whole constellation.

(Turais is visible in Britain, and is listed on page 69.)

Carina is dominated by Canopus, one of the most interesting of the really brilliant stars. It is a rival even to Sirius and is extremely luminous. Estimates of its distance and candle-power vary wildly, but it may be that Canopus is the equal of at least 80,000 Suns.

The four stars Tureis and Avior (in Carina) and Koo She and Markeb (in Vela) form what is termed the False Cross, since it is sometimes confused with the true Southern Cross (Crux Australis); but the two are easy to distinguish, since Crux is much the smaller and brighter. In Carina,

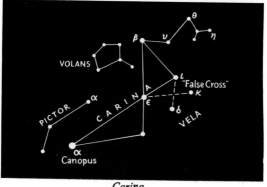

Carina

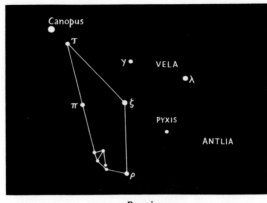

Puppis

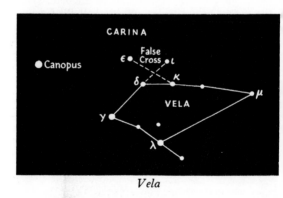

Vela

CHIEF STARS

	Visual Magnitude	*Spectrum*	*Absolute Magnitude*	*Distance, lt.-yrs.*
Carina				
α Canopus	−0·72	F0	−7·6?	650?
β Miaplacidus	1·67	A0	−0·4	86
ε Avior	1·97	K0–B	−3·1	340
ι Tureis	2·25	F0	−4·6	750
θ	2·74	B0	−4·0	710
υ	2·95	A7	−2·1	340
Puppis				
ζ Suhail Hadar	2·23	O5	−7·1	2,400
ρ Turais	2·80	F6	0·3	105
π	2·81	K4	−0·3	140
τ	2·97	K0	0·1	124
Vela				
γ	1·88	WC7	−4·1	520
δ Koo She	1·95	A0	0·2	76
λ Al Suhail al Wazn	2·24	K5	−4·6	750
κ Markeb	2·45	B2	−3·4	470
μ	2·67	B5	0·1	108

too, is the extraordinary variable star η, which is now invisible with the naked eye, but which has been known to outshine Canopus. Around 1840 it was the second brightest star in the whole sky. It is immersed in nebulosity, and it has been suggested that its fluctuations are due to 'veiling', but it must be intrinsically variable as well. It may brighten up again at any time.

γ, in Vela, is a W-type star, and is the brightest example of this rare class. Note also the Cepheid variable 1 Carinæ, which has a period of 35½ days, and changes between magnitude 3·6 and 5·0. It is extremely luminous and very remote.

The Milky Way runs through Argo, and there are many superb star-fields well shown with binoculars or a wide-field telescope.

113

PICTOR The Painter

Introduced by Lacaille in 1752, under the name of Equuleus Pictoris (the Painter's Easel). Its brightest stars are α (magnitude 3·3) and β (3·9). Pictor is notable mainly for the interesting nova, RR Pictoris, which appeared in 1925, about 7° from the north-east edge of the Nubecula Major; it reached the first magnitude, and is now a fascinating object when observed with a large telescope. Pictor lies near Canopus, and is included in the map with Carina (page 111).

VOLANS The Flying Fish

One of Bayer's additions, and formerly termed Piscis Volans. It, too, lies near Argo; in fact it intrudes into Carina, and is included with the map of that constellation. Its brightest stars—β, γ and ζ—are between the third and fourth magnitudes. γ is a wide, easy double, and ζ has a ninth-magnitude companion.

JULY STARS

During July evenings Scorpio is at its best—much more magnificent than when seen from Britain, where it is always too low to be well seen. From south Australia it is not far from the zenith, together with the neighbouring bright groups of Sagittarius and Lupus. Centaurus and the Southern Cross lie west of the zenith; Arcturus makes quite a brave show in the north, and Vega can be made out very low in the north-east. Canopus and Achernar are about at their lowest; Spica is prominent in the north-west and Altair is coming into view above the eastern horizon.

ARA The Altar

This, rather surprisingly, was one of Ptolemy's

114

	Visual Magnitude	Spectrum	Absolute Magnitude	Distance, lt.-yrs.
β	2·90	K3	−4·3	1,030
α	2·95	B2·5	−2·4	390

original groups. It is easy to find, since it is indicated by the two brilliant stars β and α Centauri.

The stars of Ara make quite a distinctive pattern, but there is little of interest in the constellation from the viewpoint of the amateur observer. It is included in the map with Triangulum Australe (page 119).

CENTAURUS The Centaur

An original constellation. It is said to represent Chiron, the wise Centaur—half-man, half-horse —who was tutor to Jason, leader of the Argonauts, as well as Hercules.

Centaurus is one of the most splendid of all the constellations, and Northerners never cease to

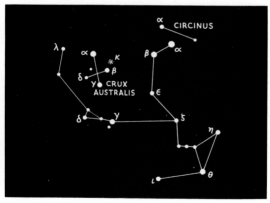

		Visual Magnitude	Spectrum	Absolute Magnitude	Distance, lt.-yrs.
α		−0·27	G2 & K1	4·4 & 5·8	4·3
β	Agena	0·63	B1	−5·2	490
θ		2·04	K0	0·9	55
γ	Menkent	2·17	A0	−0·5	160
ε		2·33	B1	−3·9	570
η		2·39 (v)	B1·5	−3·0	390
ζ		2·56	B2	−3·4	520
δ		2·59 (v)	B2	−2·7	370
ι		2·76	A2	1·1	71

regret that it is invisible from Europe. Its leaders, α and β, are close together—purely fortuitous, since α is the nearest of the bright stars (indeed Proxima, a faint member of its system, is our closest stellar neighbour) and β is a very luminous B-type giant nearly 500 light-years away. α has no recognized name; air navigators call it Rigel Kent, while β is known either as Agena or as Hadar.

α is a superb binary with a period of 80 years; Menkent is also double with a period of 80 years, though its components are much closer together. Here, too, is the globular cluster ω Centauri, the finest in the sky, and easily visible to the naked eye as a nebulous patch of about the fourth magnitude.

CIRCINUS The Compasses

Another of Lacaille's groups, added in 1752, for which there seems no real justification. Its brightest star, α, is double; the primary is of magnitude 3·4 and the companion 8·8, and the contrasting colours (yellow and reddish) make it worth looking at. Circinus lies near α Centauri, and is included in the map with Centaurus.

CORONA AUSTRALIS The Southern Crown

Corona Australis is one of Ptolemy's groups. It contains no bright stars or interesting objects, but it is perhaps worthy of a separate name, since the curved line of rather faint stars is easy to recognize.

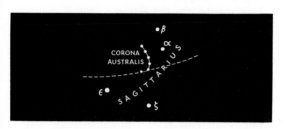

CRUX AUSTRALIS The Southern Cross

Crux is certainly the most famous of all the southern constellations. It was added to the sky-map by Royer in 1679.

CHIEF STARS

	Visual Magnitude	Spectrum	Absolute Magnitude	Distance, lt.-yrs.
α Acrux	0·87	B1 & B3	−3·9 & −3·4	370
β Mimosa	1·28	B0	−4·6	490
γ	1·69	M3	−2·5	220
δ	2·81	B2	−3·4	570

Strictly speaking, Crux is not a cross at all, but more nearly resembles a kite. It is the smallest constellation in the sky, but contains more bright stars for its area than any other, and it can hardly be mistaken. It is included here in the map with Centaurus (page 115), which more or less surrounds it, but strictly speaking it merits a separate chart!

117

Acrux is a splendid binary separable with a small telescope; the magnitudes of the individual components are 1·4 and 1·9. The red hue of γ contrasts sharply with the white of its B-type neighbours. The name Mimosa for β seems to be unofficial, but air and sea navigators use it.

Surrounding the red star κ Crucis is a superb open cluster which is known as the Jewel Box. Since the Milky Way also runs through the group, we can see that Crux is one of the richest areas in the sky.

LUPUS The Wolf

One of Ptolemy's groups. It lies between Scorpio and Centaurus.

Lupus contains several brightish stars, but it must be admitted that the group is singularly devoid of interesting objects.

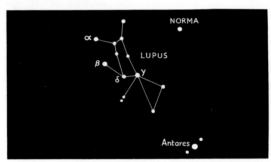

CHIEF STARS

	Visual Magnitude	Spectrum	Absolute Magnitude	Distance, lt.-yrs.
α	2·32	B1	−3·3	430
β	2·69	B2	−3·4	540
γ	2·80	B2	−2·7	570

NORMA The Rule

One of Lacaille's groups; the old name for it was Quadra Euclidis (Euclid's Quadrant). There are no bright stars, but one open cluster. N.G.C. 6067, is worth looking at with a moderate telescope. Norma is included in the map with Lupus.

TRIANGULUM AUSTRALE The Southern Triangle

A constellation added by Bayer in 1603. The name is apt, since its three chief stars do indeed form a triangle.

Triangulum Australe is easy to find, as three of its stars are quite bright, but it is otherwise unremarkable. It lies not far from α Centauri.

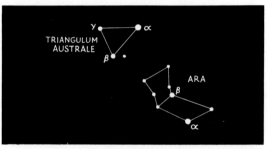

CHIEF STARS

	Visual Magnitude	Spectrum	Absolute Magnitude	Distance, lt.-yrs.
α	1·93	K2	−0·1	82
β	2·87	F2	2·3	42
γ	2·94	A0	0·2	113

SAGITTARIUS The Archer

No separate map is given here, because in theory all the brightest stars of Sagittarius rise in Britain

and they have been given in the map on page 89—
though admittedly most of them always appear so
low over the British horizon that they are difficult
to see. α and β Sagittarii, which are well below the
third magnitude, are too far south to be visible in
Europe, and this part of the constellation is
included in the map with Corona Australis (page
117).

SCORPIO The Scorpion

Here a separate map is called for, but for the sake
of completeness the rest of the constellation has
also been included. The 'sting' of Scorpio never
rises over Britain, but is very prominent. The
bright stars never visible from Britain are:

		Visual *Magnitude*	*Spectrum*	*Absolute* *Magnitude*	*Distance,* *lt.-yrs.*
λ	Shaula	1·60	B1	−3·3	310
θ	Sargas	1·86	F0	−4·6	650
κ	Girtab	2·39	B2	−3·4	470
υ	Lesath	2·71	B2	−3·4	540
ι′		2·99	F2	−7·1	3,400

Shaula, Lesath, Girtab and G (magnitude 3·25)
make up the 'sting'. The whole area is rich, and

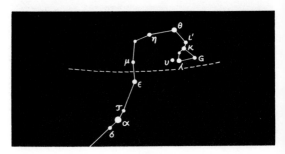

120

there are many splendid telescopic star-fields. Moreover Scorpio, when seen in its entirety, is one of the few constellations which really does look a little like the creature it is meant to represent!

OCTOBER STARS

In October the Southern Birds are at their best, not far from the zenith: the Crane, the Peacock, the Phœnix, the Toucan—all quite prominent. Fomalhaut is also very high, and the Square of Pegasus may be seen in the north, with Altair in the west; Deneb is visible very low in the north-west, and Vega too may be glimpsed soon after sunset. Scorpio is descending in the west; Canopus may be seen in the south-east, and the Southern Cross is almost at its lowest. Achernar, high up, is east of the overhead point. Much of the northern aspect is occupied by the large, dull groups of Cetus, Capricornus and Aquarius.

GRUS The Crane

Grus is the most prominent of the Southern Birds. Like the Peacock, the Phœnix and the Toucan, it was added by Bayer in 1603.

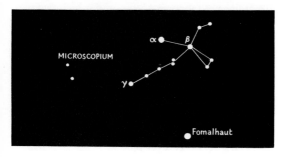

CHIEF STARS

		Visual Magnitude	Spectrum	Absolute Magnitude	Distance, lt.-yrs.
α	Alnair	1·76	B5	0·3	64
β	Al Dhanab	2·17 (v)	M3	−2·5	280
γ		3·03	B8	−3·1	540

Grus lies near Fomalhaut, and does give some impression of a flying crane—provided that the observer draws liberally upon his imagination. Alnair is bright enough to be very prominent, and the group is easy to find, but it does not contain many objects of note.

INDUS The Indian

Another of Bayer's groups. It lies close to α Pavonis, and is included in the map with Pavo (page 123). ε, magnitude 4·7, lies at a distance of only 11·4 light-years, and is therefore one of our nearest stellar neighbours. The brightest star in Indus is α, magnitude 3·2.

MICROSCOPIUM The Microscope

An entirely unremarkable group introduced by Lacaille. It has no star brighter than magnitude 4·7. It lies not far from Alnair, and is included with the map of Grus (page 121).

PAVO The Peacock

One of the more conspicuous Birds. It is easy to identify, as it contains several brightish stars. It borders on Ara and Triangulum Australe.

CHIEF STAR

	Visual Magnitude	Spectrum	Absolute Magnitude	Distance, lt.-yrs.
α	1·95	B3	−2·9	310

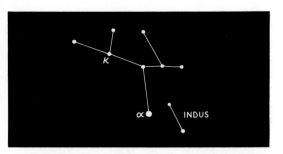

The most interesting object in Pavo is the Cepheid variable κ, which varies between magnitudes 4·0 and 5·5 in a period of 9·1 days. Its fluctuations may be followed with the naked eye.

PHŒNIX The Phœnix

A mythological bird, which was regularly consumed by fire and rose again out of its ashes.

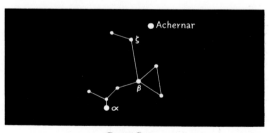

CHIEF STAR

	Visual Magnitude	Spectrum	Absolute Magnitude	Distance, lt.-yrs.
α Ankaa	2·39	K0	0·1	93

Phœnix lies near Achernar; as well as Ankaa it contains two more stars, β and γ, above the fourth magnitude. β is a good, rather close binary, while ζ, of the fourth magnitude, has a companion of magnitude 8·4.

TUCANA The Toucan

Tucana is the faintest of the Birds, but probably the most interesting in view of the nebular objects which it contains.

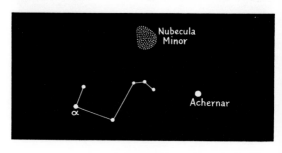

	Visual Magnitude	Spectrum	Absolute Magnitude	Distance, lt.-yrs.
α	2·87	K3	1·5	62

Most of the Nubecula Minor, or Small Magellanic Cloud, lies in Tucana; here also are two globular clusters—47 Tucanæ, which ranks second only to ω Centauri and is prominently visible with the naked eye, and N.G.C. 362, which may just be glimpsed without optical aid as a blur of about the sixth magnitude. These three objects more than make up for the paucity of bright stars in the constellation.

This description of the constellations is naturally very rough and incomplete, but at least it may serve to show the variety of objects which may be seen with modest telescopes—or, indeed, with no telescopes at all. When the clouds roll back and the stars shine down, the night sky can never be dull.

THE STELLAR UNIVERSE

Anybody who practises for a few weeks will soon learn how to recognize the different stars and constellations. The amateur with a small or moderate telescope may, if he wishes, make himself useful by estimating the magnitudes of variable stars, and possibly by measuring the position angles and separations of binary pairs. This is about as far as most people will go; to become a research astrophysicist entails a thorough mathematical training, and most of the work is done at a desk instead of at the eye-end of a telescope. However, it may prove of interest to add a few remarks about modern astrophysical theories.

We have seen that the stars are made up of incandescent gas, and that their surface temperatures must be measured in thousands of degrees. Obviously they must be still hotter inside, and they must be able to draw upon immense reserves of energy. One of the greatest problems facing astronomers of the early years of the present century was: What keeps the stars shining?

Several important points had to be borne in mind. It was found, for instance, that the red stars were divided into two well-marked groups: the *giants*, of vast size and tremendous candlepower, and the *dwarfs*, relatively small, and much feebler than the Sun. For instance, both Betelgeux in Orion and our nearest stellar neighbour Proxima Centauri show M-type spectra, and have much the same surface temperature; but there the resemblance ends. Betelgeux is a typical red giant, while Proxima is a red dwarf. If we

126

represent Betelgeux by a searchlight, Proxima will be a match.

The giant and dwarf division was also evident with the orange stars, and with the yellow, though in the latter case the discrepancy between the two groups was not so great. There was no such division with the white stars, all of which appeared to be hot and highly luminous.

If we plot the stars according to their spectral type and luminosity, as is done on the famous *Hertzsprung-Russell Diagram*—so called because it was first worked out in detail by the Danish astronomer Ejnar Hertzsprung and H. N. Russell of America—the pattern is striking, and led Russell to put forward a theory of stellar evolution which was accepted for many years. It was a modification of an earlier scheme by Sir Norman Lockyer, a British astrophysicist, but was a great step forward, because Russell made a serious attempt to solve the problem of the energy source of the stars.

First, it is obvious that the stars cannot be 'burning' in the accepted sense of the word. A body the size of the Sun, made entirely of coal and radiating as fiercely as the Sun actually does, could not last for more than a few million years, and we can prove that the Sun's age must be reckoned in thousands of millions of years—simply because the Earth itself can be shown to be about 4,700,000,000 years old. So the Sun and other stars must have definite ways of producing enough energy to last them for such immense spans of time.

Russell suggested that this energy might be drawn from the annihilation of matter. All the matter in the universe is made up of a comparatively few fundamental substances or *elements*; ninety-two are known to occur naturally, among which

127

are familiar solids, liquids and gases such as iron, mercury and hydrogen. Each *atom* is composed of a nucleus around which revolve numbers of *electrons*—one electron for the hydrogen atom, two for helium, three for lithium, and so on up to ninety-two for uranium. On Russell's theory an electron, which carries a unit negative charge of electricity, could cancel out a *proton*, another fundamental particle, this time carrying a unit positive charge of electricity. Both particles would then disappear, with the production of a certain amount of radiation.

Russell considered that a star would begin its career by condensing out of interstellar dust and gas. At first it would be a large, cool red giant; but as it shrank, under gravitational forces, it would become denser and hotter, eventually turning into a very luminous white star of spectrum B or O. Subsequently it would continue to shrink, but would cool down, ending its active career as a faint red dwarf. Its final fate would be the loss of all its light and heat, so that it would become a dead globe.

Russell's theory was accepted for many years, but modern research has shown that it is incorrect, plausible though it may seem. The stars shine not because of the direct annihilation of electrons and protons, but because of nuclear reactions deep inside their globes.

Consider first an ordinary star such as the Sun, which consists largely of hydrogen—the most abundant element in the universe. Near the centre of the solar globe the temperature must attain 14,000,000°C, and strange things are happening to the hydrogen nuclei. They are joining together to form nuclei of helium; four of them are needed to make one helium nucleus, but each time the process

128

occurs there is a certain loss of mass, together with the emission of energy.

What is happening, then, is that the Sun is changing its hydrogen into helium. It is losing mass at the rate of 4,000,000 tons per second and yet the energy available is sufficient to last it for several thousand millions years to come. Eventually the hydrogen 'fuel' will be used up and radical changes will take place in the structure of the Sun, but at present there is no fear of any marked alteration in the output of light and heat.

According to current ideas a star begins—as Russell supposed—by condensing out of interstellar material. The gaseous nebulæ, both bright and dark, are regarded as birthplaces of the stars, and it is highly probable that fresh stars are being produced in nebulæ such as the Sword of Orion. The embryo star shrinks by gravitational contraction; when the interior temperature has risen sufficiently, nuclear reactions begin, and the star starts to shine.

Russell supposed that a normal or *main sequence* star, such as the Sun, must be cooling down. Actually this seems not to be the case; as it ages, and uses up its hydrogen, the Sun is becoming more luminous. When its nuclear fuel is almost exhausted, as may be the case in—say— 6,000,000,000 years from now, the outer layers will swell and cool down, so that the Sun will turn into a red giant. The effects of this change will certainly prove disastrous for the Earth, but in any case we may safely assume that the temperature on our world will have become intolerably high even before the Sun reaches the red giant stage.

Subsequently it is thought possible that the Sun will 'collapse', rather abruptly on the astronomical scale, into a very small, very dense star known as a white dwarf. Many such white dwarfs are known, the most famous example being the faint companion of Sirius. In them the atoms are broken up, so that the component nuclei and electrons are jammed together with little waste of space; the material is unbelievably dense, and a matchboxful of white dwarf material would weigh many tons. These stars have been nicknamed 'stellar bankrupts'. They have squandered all their nuclear fuel, and shine feebly only because they are still contracting slowly under gravitational forces. In time the last of their energy will leave them.

A more massive star may explode as a supernova, leaving a patch of expanding gas; the Crab Nebula is such a remnant. In it is a pulsar, which is a quickly varying radio source, and may be what is called a neutron star, of small size and almost incredible density. An even more massive star may collapse so that not even light can escape from it, and it will become what is termed a 'black hole'.

There is no shortage of dust and gas in the Galaxy. In addition to the bright nebulæ such as the Sword of Orion, and the dark nebulæ which block out the light of stars beyond, there is material scattered all through the system, and there can be little doubt that this is the new material out of which stars are born.

Sir William Herschel, the 'father of stellar astronomy', drew up the first reasonably accurate picture of the shape of the Galaxy, and suggested that the form was similar to that of a double-convex lens. He was wrong, however, in placing the Sun near the centre. In reality, the Sun is about 32,000 light-years from the galactic nucleus, so that it lies well out toward one edge of the system.

The key to the problem was provided by the globular clusters, of which the brightest example visible from Britain is Messier 13 in Hercules. Soon after the end of the First World War, the American astronomer Harlow Shapley studied the distribution of the globulars, and noted that most of them lie in the southern part of the sky—in the region of Scorpio and Sagittarius. Such an unsymmetrical arrangement could hardly be coincidental, but at that time it was impossible to make any accurate distance measures, since all the globulars are very remote. Fortunately some interesting variable stars came to the rescue. These were the Cepheids, named after the prototype star δ Cephei.

The importance of these variables lies in the fact that there is a definite law linking their light-changes with their real luminosities. For example, δ Cephei has a period of 5 days 9 hours; any other Cepheid with a period of 5 days 9 hours must be intrinsically as luminous as δ Cephei itself. If the period is longer, as with η Aquilæ (just over 7 days), the luminosity is greater; if the period is less, the luminosity also is less. We can therefore work out the candlepower of a Cepheid simply by watching its fluctuations; and this gives us the distance. The reason for the period-luminosity law is unknown, but the law is undoubtedly valid.

Related to the Cepheids are the so-called RR Lyræ stars (because the star lettered RR in Lyra is the best-known member of the class), which have shorter periods, and all of which are about 90 times as luminous as the Sun. Shapley was able to detect RR Lyræ stars in some of the globular clusters, and thus to work out the distances of the clusters in which the variables lay. Since the globulars form a sort of outer framework to the Galaxy, he was thus able to draw up a really reliable picture of the whole system. It proved to be about 100,000 light-years from side to side, with a maximum thickness of some 20,000 light-years.

It has also been found that the Galaxy is rotating around its nucleus; the Sun takes about 225,000,000 years to complete one circuit. Unfortunately the interstellar material is densest in the main plane of the system, and effectively blocks out the actual centre. The Sagittarius star-clouds give us the directions, but we can never have a direct view.

In addition to emitting visible light, celestial bodies may send out radiations of longer wave-length, known as *radio waves*. The name is rather misleading, since there is no suggestion that they are artificial transmissions, and of course they cannot be seen. They may, however, be collected by special instruments known as *radio telescopes*, which are really in the nature of large aerials. The first such instrument was built by Karl Jansky, an American engineer, in the early 1930s; the most famous in the world today, situated at Jodrell Bank in Cheshire, has a 'dish' 250 feet in diameter.

Radio waves are not blocked by interstellar matter, and have given us the means of exploring those parts of the Galaxy which can never be examined optically. It has been found, too, that the Galaxy is spiral in form, not unlike an immense, rather loosely wound Catherine wheel. The sun appears to lie near the edge of one of the spiral arms.

This is not so unexpected as might be thought. Far beyond our own Galaxy can be seen other systems, many of which are spiral. Of these the most famous is the Great Galaxy in Andromeda, which is dimly visible to the naked eye on a clear night. Large instruments show its form quite clearly, though in a small telescope it is frankly a disappointment.

Until the present century it was widely believed that the spirals and other such objects belonged to our system, but in 1923 E. P. Hubble, at Mount Wilson Observatory in California, managed to find Cepheid variables in some of them. By observing the Cepheids he was able to draw up a reliable distance scale, and his results were conclusive. Far from being a member of our Galaxy, the Andromeda Spiral was shown to be a galaxy in its own right. It lies about 2,200,000 light-years from us, and is a system even larger than ours. It too contains its quota of stars, nebulæ and clusters; even novæ and supernovæ have been seen in it from time to time.

The world's most famous optical telescope, the Palomar 200-inch reflector, can photograph about 1,000,000,000 galaxies. Since each contains an average of perhaps 100,000 million stars, we can see that the number of 'suns' in the observable universe is truly colossal. The logical—though unproved—inference is that planetary systems also must be common, and it is likely that inhabited worlds are widespread, though naturally we cannot be sure.

The most remarkable thing about the galaxies is that apart from those in our local group, including less than thirty members, all of them seem to be racing away from us at tremendous velocities, with the speeds of recession increasing with distance. Though the remote galaxies are tremendous systems, they appear on our photographic plates as mere specks of light.

There is no suggestion that our Galaxy is particularly unpopular; if the velocity measures are to be believed, the entire universe is expanding. How this expansion began is a matter for conjecture. On one theory the universe began at a definite moment in the past, and will eventually die; on another, the universe has always existed, and goes through alternate phases of expansion and contraction.

The problem has been made still more complicated by the discovery in 1963 of an entirely new class of objects. These objects look very much like faint stars, but their spectra are quite different, and they have been called quasi-stellar objects or, for short, quasars. Present evidence indicates that they are immensely luminous and immensely remote—even further away than the most distant 'normal' galaxies—and since they are relatively small, they must be producing their energy in a way which we do not so far understand. Quasars are probably the centres of very active galaxies.

Problems of this sort must be left to the highly trained professional research worker, and the amateur observer can do nothing to help. Few of the outer galaxies are to be seen with small telescopes, and even the Great Spiral in Andromeda is not spectacular. Yet the problems themselves are perhaps the most fascinating known to us, and we can only hope that before many years have passed we shall be able to unravel at least some of the riddles of these remote star-systems.

THE NEAREST STAR—THE SUN

Astronomy is generally regarded as one of the safest of hobbies. Unless an observer falls through the roof of his observatory, or catches cold through staying outdoors for too long on a bitter night, it is not obvious how he can come to any harm. Yet there is one real danger facing the unwary beginner; he may damage his eyes by looking at the Sun through a telescope.

The Sun is a normal star, but it is far closer to us than any of the rest. The astronomical unit (Earth–Sun distance) measures 93,000,000 miles, which is not a great deal in comparison with the vast stretches of space which separate the Solar System from its neighbours. Consequently the Sun sends us a tremendous amount of light and heat; and to concentrate this heat on to one's eye is fatal.

Most people remember the old Boy Scout method of starting a camp-fire. When the Sun is available, all that need be done is to use a piece of glass, such as a spectacle lens, to focus the heat on to dry plant-stuff. Almost at once the material takes fire. (It is also, unhappily, true that many forest fires are started by careless holidaymakers who omit to clear up after a picnic, and leave broken pieces of glass lying around to concentrate the solar rays.)

The object-glass of a refractor, or the mirror of a reflector, concentrates the Sun's heat in just the same way. If this heat is focused on to the eye, the result will be permanent blindness. Many beginners have injured themselves seriously in this way; I have personally come across several such cases.

The danger cannot be over-stressed. Even when the Sun is low in the sky, and seems gentle and harmless, there is still a grave risk to anyone who looks at it through binoculars or a telescope. A second's view may be too much.

Unfortunately, it is possible to buy dark 'sun-caps' which are made to fit over the eyepiece of the telescope. It is then—so we are told—safe to look directly at the Sun. What we are *not* told is that

135

sun-caps are always liable to splinter without warning, and will in any case give inadequate protection. I have frequently urged that the sale of all such dark caps should be stopped; certainly they should never be used.

However, there is one perfectly safe way to look at the Sun, and this is by the method of projection. A refractor with an object-glass of 3 or 4 inches' diameter is ideal; a 5-inch is too large, since it collects more light (and heat) than is desirable. Reflectors are less suitable, unless the mirror is left unsilvered—which naturally makes it useless for any other sort of work. Though I have made daily observations of the Sun for many years now, I have never yet turned my main reflector toward it; instead I use an old 4-inch refractor.

The first step is to point the telescope at the Sun, keeping an opaque cap over the object-glass. Then remove the cover and project the solar image on to a sheet of white paper or card held behind the eyepiece. After minor adjustments to the focus the Sun should be excellently seen; my 4-inch refractor gives a sharp picture over one foot across.

Various refinements may be added. It is as well to fix a screen over the telescope to cast shadow on to the projection-sheet, and there is little difficulty in making a framework or box to hold the screen itself. Moreover, various eyepieces may be used to give different magnifications.

A smaller telescope, such as a home-made spectacle-lens refractor, is naturally less efficient, and the definition will be inferior. However, a properly mounted instrument of this kind will suffice to show the most interesting features of the solar disk, the darkish patches known as *sunspots*.

A sunspot appears as a blackish area against the bright surface or *photosphere*. The darkness is

136

deceptive, and is an effect of contrast. Whereas the photosphere has a temperature of about 6,000° Centigrade, the spot is 2,000° cooler, and so emits less light; yet if it could be seen shining by itself its brilliancy, area for area, would far surpass that of an arc lamp.

A large spot is made up of a central portion or *umbra* surrounded by a lighter area or *penumbra*. Generally speaking the outline will be to some extent irregular; perfectly circular spots are not uncommon, but others are highly complex, and many umbræ may be contained in a single area of penumbra. We must, moreover, reckon with the effects of foreshortening. A spot near the Sun's limb will be seen 'at an angle'; if its true shape is circular, it will appear to be drawn out into a long, narrow ellipse.

Spots may appear singly, but more often form groups. It is a common sight to see two main spots, one lying to the west of the other, with numerous smaller spots near by. A major group may contain dozens of separate umbræ, and sometimes the detail is so complex that it is almost impossible to sketch accurately.

Spots are comparatively short-lived. Smaller ones may last for less than a day, while even large groups persist for only a few weeks or months. Their apparent movements are interesting to follow. Since the Sun rotates on its axis in a period of slightly less than a month, the spots seem to be carried across the disk from one limb to the other, and the daily shift is very evident. When a spot vanishes over the limb it is naturally lost to view, but if it persists it will reappear at the opposite limb about a fortnight later. Occasional groups survive long enough to cross the Sun several times.

The spots appear small, but in reality they cover vast areas; the largest group ever recorded—that of April 1947—extended over a total of at least 7,000,000,000 square miles. It is not uncommon for a spot to become large enough to be seen with the naked eye when the Sun is veiled by mist or haze, and records of naked-eye spots go back to the days of the ancient Chinese civilization.

Sometimes the Sun is seen to show numerous spot-groups at the same moment, while at others the disk is completely free from them. It has been found that the spot-frequency varies in a definite cycle of about 11 years. Maxima occurred, for instance in 1968–69 and in 1979–80, so that the next may be expected about 1991—though it is impossible to be precise, since the cycle is by no means perfectly regular. Midway between the maxima, as in 1974–75, the disk is much less active, and it may be that no spots will be seen for days on end. We must also expect only low solar activity in the mid-1980s.

We know what sunspots are, and we know how they behave; they alter in form relatively quickly, and a large spot-group may be so complex that it is difficult to sketch accurately. Photographs are taken each day at professional observatories, and it is idle to pretend that the modestly equipped amateur can make major contributions to solar work, but it is well worth while to keep track of the spots and watch them as they shift and change.

Associated with spots are bright irregular patches known as *faculæ*, from the Latin word meaning 'torches'. Faculæ lie well above the photosphere, and may be described as luminous 'clouds' hanging in the upper regions. They often appear in an area where a spot-group is about to break out, and persist for some time after a group

138

has disappeared. Projection with a small refractor will show the faculæ very well.

There are many problems connected with sunspots and faculæ. It is known, for instance, that the spots have strong magnetic fields, and send out particles which strike the Earth's upper atmosphere, producing the wonderful displays of polar lights or aurorae. Research with more complicated equipment, based on the principle of the spectroscope, can yield valuable and fascinating results, but unfortunately few amateurs can hope to acquire instruments of this kind. Incidentally, it used to be thought that sunspots caused bad weather, and it is certainly true that the spot-maxima of 1938–39 and 1947 coincided with very cold winters in Britain; but nowadays this supposed connection is generally rejected by astronomers.

The great brilliance of the photosphere means that the outer atmosphere of the Sun cannot be normally seen, except with the use of complex instruments. There are occasions, however, when we are privileged to have an unrestricted view. Our best moments for seeing the Sun in its full glory come during total solar eclipses.

The Moon is much smaller than the Sun—its diameter is a mere 2,160 miles, as against the 865,000 miles of the Sun—but it is so much closer to us that it appears almost exactly the same size. When the three bodies move into a direct line, therefore, the disk of the Moon just covers the

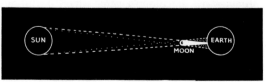

solar photosphere, and the normal glare is cut out, so that the Sun's outer surroundings flash into view.

The diagram shows the theory of a solar eclipse. The shadow of the Moon touches the Earth, and from points lying inside the zone of totality the Sun is completely hidden. Unfortunately this zone is narrow, and can never exceed a width of 167 miles. To either side there is an area on the Earth from which the Sun will be partly blotted out by the Moon, and a *partial eclipse* will be seen. It is clear that a solar eclipse can happen only at the time of new moon, when the dark part of the lunar disk is turned in our direction and the Moon itself cannot be seen directly.

A third kind of eclipse is the *annular*. The Moon's distance from the Earth varies within certain limits, and when exact lining-up occurs when the Moon is at its most remote (about 252,000 miles from us) the lunar disk is too small to cover the Sun completely. Consequently a narrow ring of the photosphere is left showing round the dark body of the Moon.

Partial and annular eclipses are interesting to watch, but are of little importance to the astronomer; even the slightest crescent of the photosphere is enough to conceal the Sun's atmosphere. When the eclipse is total, however, matters are very different. As soon as the Moon has drawn completely over the Sun, the sky becomes dark enough for stars to be visible; and we see both the *chromosphere* or colour-sphere, which lies above the solar photosphere, and the outer *corona*.

The corona is particularly splendid. It forms the outermost part of the Sun's atmosphere, and is composed of very tenuous gas at a remarkably high temperature. Sometimes well-marked streamers are seen stretching across the sky, while at other

140

eclipses the outline of the corona is more regular. It extends for many millions of miles, though it is so tenuous that it can have no definite boundary.

Also visible at some eclipses are the *prominences*, still often called by their old but misleading name of 'red flames'. They too are made up of incandescent gas, and are of tremendous size, since the length of an average prominence amounts to 125,000 miles. Some of them are *eruptive*, and relatively quick-moving; others, the *quiescent* prominences, may last for weeks before breaking up gradually or being violently disrupted.

Nowadays special instruments have been built which enable astronomers to study the chromosphere and prominences at any time, together with the innermost part of the corona; but only during total eclipses may they be seen with the naked eye, and it is not surprising that full use is made of our limited opportunities. It is worth noting, too, that there are certain specialized observations which can be made only during a total eclipse.

The last total eclipse visible in Britain took place in 1927 (though the eclipse of 30 June 1954 was just total off the north Scottish islands). The next will not occur until 11 August 1999, though other parts of the world are more favoured. The next partial eclipses visible from Britain will occur on 30 May 1984, 10 May 1994 and 12 October 1996.

Solar eclipses caused great alarm in ancient times; the Chinese used to believe that the Sun was in danger of being eaten by a dragon. No terror is now associated with them, except in very undeveloped countries, but they remain perhaps the most awe-inspiring phenomena in all nature. Nobody who has been fortunate enough to witness a total eclipse of the Sun is ever likely to forget it.

AURORÆ, THE ZODIACAL LIGHT AND THE GEGENSCHEIN

Before turning to the other bodies of the Solar System, it will be as well to deal briefly with the various 'glows in the sky' which are seen from time to time. First must come the polar lights, or aur:, since their origin lies in the Sun.

Certain active regions of the solar disk—frequently, though not invariably, associated with large spot-groups—send out electrified particles. These particles cross the 93,000,000-mile gap between the Sun and ourselves and strike the Earth's upper atmosphere, causing aurorae. The heights of the displays vary between over 600 miles above ground level down to as little as 60 miles. Sometimes the aurorae take the forms of regular patterns, while at others they shift and change rapidly, producing effects which are indescribably beautiful. The colours, too, may be spectacular. The brightest British aurora of modern times, that of 25 January 1938, was vivid red in hue, and some people in southern England believed all London to be on fire.

Since the particles are magnetic, they are drawn to the magnetic poles, though the Van Allen zones surrounding the Earth are known to be closely involved. This is why aurorae are best seen in high latitudes. During the long winter nights in northern Norway, auroral displays are seen almost constantly; in Scotland there are generally a good many bright displays each year, but in southern England aurorae are rare. City-dwellers in the Home Counties, for instance, will almost never see them.

In the north the phenomenon is called Aurora Borealis (the Northern Light), while in the Antarctic the term Aurora Australis (Southern Light) is used.

A typical display may begin as a glow on the horizon, rising gradually to become an arc. The bottom of the arc then brightens, sending forth streamers, after which the arc loses its regularity and develops 'folds' in the manner of a shining curtain. If the streamers extend beyond the zenith they converge in a patch termed an *auroral corona*. Finally the display sends waves of light from the horizon toward the zenith, after which the glow dies slowly away. The whole phenomenon may extend over many hours.

No instruments are needed for watching auroræ, and the main requirement is to observe well away from artificial lights. Unfortunately it is impossible to predict just when a display will be visible, but a rough guide may be obtained from the state of the Sun; when the solar disk is particularly active, auroræ will be more frequent than when spots are rare. On the other hand, there is no guarantee that any particular large spot-group will produce an aurora.

Much less spectacular than auroræ is the Zodiacal Light, which takes the form of a cone-shaped radiance seen after dusk or before dawn, extending upward along the ecliptic from the hidden Sun. From Britain it is never prominent, though it may become conspicuous from countries where the atmosphere is more transparent and less dust-laden. It is best seen in late evening in March and early morning in September, since at these times the ecliptic makes its steepest angle to the horizon.

The Zodiacal Light has no association with

aurøræ. It is due to a layer of thinly spread matter in the main plane of the Solar System, stretching from near the Sun as far out as the Earth's orbit—and probably well beyond. At its best its brightness is comparable with that of the Milky Way.

Third and last of these lights in the sky is the Gegenschein or Counterglow, described in 1854 by the Danish astronomer Theodor Brorsen. It is visible exactly opposite to the Sun in the sky, and is best seen in September, when it looks like a round luminous patch about 40 times the apparent width of the Moon. The Gegenschein is very elusive, and I have seen it from Britain only once—in 1940, when the whole country was blacked out because of the danger of air-raids, and conditions for astronomical observation were consequently very good. The Gegenschein also is due to thinly-spread interplanetary material; like the Zodiacal Light, it originates well beyond the limits of the Earth's atmosphere.

Amateurs may make themselves very useful in recording the appearance and frequency of aurøræ and the Zodiacal Light, and indeed a well-organized programme carried through during the period of the International Geophysical Year produced valuable results. However, it must be realized that people who live in the northernmost parts of the British Isles have the best views, and the opportunities for dwellers in southern England are regrettably restricted.

THE MOON

The Moon is the Earth's only natural satellite. It is a very minor body, but its closeness to us makes it appear far more splendid than any object in the sky apart from the Sun, and so far it is the only world which has been reached by Man. Its diameter is 2,160 miles, and its average distance from the Earth is slightly less than 239,000 miles.

The monthly phases are easily explained. Since the Moon has no light of its own, and depends solely upon reflecting the light of the Sun, one hemisphere is illuminated while the other hemisphere is dark. When the Moon is approximately between the Earth and the Sun, its dark side is turned in our direction; the Moon is the 'new', and cannot be seen at all. (Many people term the crescent moon 'new', but this is astronomically incorrect.) As the Moon moves round the Earth, part of the daylight side begins to face us; the phase grows from crescent successively to half, three-quarters (*gibbous*) and then full, when the whole of the lighted hemisphere may be seen. During the next fortnight the phases are repeated in the reverse order, until the Moon is again 'new'.

One or two qualifications should be made here. It is not strictly correct to say simply that the Moon revolves round the Earth, since what actually happens is that the two bodies revolve round their common centre of gravity. However, the Earth is 81 times as massive as the Moon, and the centre of gravity of the Earth–Moon system lies inside the terrestrial globe, so that the simplified statement is good enough for most purposes.

Secondly, it is clear that when the Moon is exactly between the Earth and the Sun at the moment of new moon, there will be a solar eclipse. This does not happen every month, since the lunar orbit is tilted with respect to that of the Earth, and generally the Moon passes either above or below the Sun in the sky.

145

The Earth casts a shadow in space, and it some-times happens that the Moon passes into this shadow. The result is a lunar eclipse, which may be either total or partial. Ordinarily the Moon does not vanish altogether, since the layer of atmo-sphere round the Earth refracts a certain amount of sunlight on to the lunar surface; the Moon becomes dim, and at some eclipses the colouring is beautiful. A future lunar eclipse visible in Britain will take place on 4 May 1985 and will be a total eclipse. For obvious reasons it is clear that there can be no such thing as an annular eclipse of the Moon; moreover, equally obviously, a lunar eclipse can take place only at full moon.

When lunar eclipses occur, they may be seen over a complete hemisphere of the Earth. They are far less important than total eclipses of the Sun, but they are lovely to watch; it is fascinating to see the Earth's shadow creeping gradually across the face of the Moon.

It may also happen that in its journey across the sky the Moon passes over, and *occults*, a star—or, occasionally, a planet. The star seems to shine steadily until it is covered by the lunar limb, when it snaps out abruptly; the immersion of a planet naturally takes some seconds, since the planet shows a perceptible disk. When an occultation takes place at the dark limb of the Moon, the effect is spectacular. Occultations of bright stars and planets may be well seen with binoculars, and it is a pity that they are rather uncommon. A 3-inch refractor is powerful enough to give good views of the occultations of fainter stars.

Even with the naked eye, dark and bright areas may be seen on the Moon's disk. Near full, these markings are said to give a rough impression of a human face; who has not heard of the Man in the

146

Moon? But when optical aid is used, the Man is lost in a maze of intricate detail.

The Moon is a strange, unfriendly world. It has no atmosphere, and so its surface features appear sharp and clear-cut. Any small telescope will show the dark plains which are known as maria or *seas*, as well as mountain ranges, valleys, and hundreds upon hundreds of walled circular formations which are termed *craters*. There is no water anywhere on the Moon, and the so-called seas are dry tracts without a scrap of moisture in them; but they were once thought to be liquid, and the old names are still used, generally in the Latin form. We have, for instance, the Mare Nubium (Sea of Clouds), Oceanus Procellarum (Ocean of Storms), Sinus Iridum (Bay of Rainbows) and many more. Some of them are of great extent; the Mare Imbrium (Sea of Showers) is as large as Great Britain and France put together, while the Oceanus Procellarum is considerably bigger than our Mediterranean.

The mountains are high—the loftiest of all may attain almost 30,000 feet—and isolated peaks and clumps of hills are also common. But the lunar scene is dominated by the craters, ranging in size from vast enclosures well over 150 miles across down to tiny pits. Some of them possess high central mountains or mountain-masses, while others have comparatively smooth floors. Most of them are named after scientists of the past or present, though a few mythological gods and heroes have found their way to the Moon (notably Hercules and Atlas), together with a few famous historical characters such as Julius Cæsar and Alexander the Great.

A typical large crater has a rampart which rises to only a moderate height above the outer surface;

the slope of the wall is not so steep as might be thought. The deepest part of the floor may be anything up to 30,000 feet below the crest of the rampart, but it is important to remember that a lunar crater is not really deep in proportion to its diameter. Its profile resembles a shallow saucer more than a deep well.

When a crater is near the *terminator*, or boundary between the lighted and unlighted hemispheres of the Moon, its interior is largely filled with shadow, and the crater is very conspicuous. As the Sun rises higher over it, the shadow decreases and finally disappears, so that the crater loses its prominence. Oddly enough, full-moon is the very worst time to start an observational programme. At such times the sunlight is striking directly on to the Moon; there are few shadows and most of the craters are difficult to identify. The best periods for observation are around half-moon.

The origin of the craters remains something of a puzzle. On one theory they are due to meteoric impacts, in which case they must be relations of terrestrial meteor craters such as Coon Butte in Arizona. This idea sounds plausible enough, but there are strong objections to it, and many authorities believe that the craters are basically volcanic. No doubt both processes have operated.

Mild activity may be in progress even now, and reddish patches lasting for relatively short periods have been seen in and near various craters—for instance, Alphonsus (N. A. Kozyrev, 1958); Aristarchus (J. C. Greenacre and others, 1963); Gassendi (P. Sartory, T. J. C. A. Moseley, P. Ringsdore and myself, 1966). It seems that these transient lunar phenomena, often nicknamed T.L.P.s, are less rare than has been thought, though they are, of course, unpredictable and elusive.

Various craters appear to be the centres of systems of bright streaks or *rays*, which extend for great distances across the Moon. These rays are surface deposits, crossing all other surface features, and are well seen only under high light. They are very obscure until after half-moon, but near full dominate the picture. The two main systems are associated with the craters Tycho, in the southern part of the Moon, and Copernicus, in the Oceanus Procellarum.

At this point it may be useful to give some notes about the chief lunar features visible with small telescopes, starting at new-moon, the beginning of the *lunation*, and working through to full-moon. The maps are drawn with north at the bottom, in accordance with the usual astronomical practice; those who are using ordinary binoculars will have to turn the charts upside-down.

1st, 2nd and 3rd days after New Moon. The Moon is a thin crescent, and very often the "dark" side may be seen shining faintly. This appearance is known commonly as 'the Old Moon in the New Moon's arms" but its official name is the Earth-shine, since it is due to light reflected by the Earth on to the Moon.

By Day 3, the terminator passes across a well-defined dark sea, the Mare Crisium (Sea of Crises), which measures 280 miles by 350, thus enclosing an area of 66,000 square miles—larger than England. Well south of the Mare Crisium are several large craters, notably Langrenus, Vende-linus and Petavius. All these are over 80 miles in diameter, but they are not alike; Vendelinus is less regular than the other two, and seems to have been partly ruined, so that it is presumably older. Petavius is notable for the great *cleft* which runs from the central mountain to the south-east wall. A small telescope will show it.

4th Day after New Moon. The crescent is broader, and Mare Crisium is completely visible. Closely east of it is a very brilliant crater, Proclus, 18 miles in diameter and 8,000 feet deep; it is the centre of a system of rays, some of which cross the Mare Crisium. Langrenus, Vendelinus and Petavius are still prominent, and so is the 78-mile crater Endymion in the north, which is easily recognizable because of its dark grey floor. Another very important feature is the large sea known as the Mare Fœcunditatis (Sea of Fertility).

5th Day after New Moon. The most striking features are the three great craters Theophilus, Cyrillus and Catharina which form a chain with Cyrillus in the middle. They lie on the eastern border of a well-marked sea, the Mare Nectaris (Sea of Nectar), and are very prominent indeed, with shadow covering most of their floors. Theophilus has a diameter of 64 miles, with terraced walls rising to 18,000 feet above the interior; it is one of the grandest craters on the whole Moon. Cyrillus and Catharina are similar in size, but are less deep.

The Mare Crisium and the Mare Fœcunditaris remain conspicuous, and we can now see much of another large plain, the Mare Tranquillitatis (Sea of Tranquillity). Note also the large craters Posidonius and Piccolomini; Piccolomini marks the western end of a famous scarp, the Altai Scarp. The southern part of the Moon is particularly rough, and craters occur in great numbers.

6th Day after New Moon. We can now see all of the Mare Tranquillitatis, and most of its even more clearly marked neighbour the Mare Serenitatis (Sea of Serenity). There are few craters on the Mare Serenitatis, the only conspicuous one being Bessel, a dozen miles across. One very interesting feature is the white patch known as Linné. Before 1843 it was described as a prominent crater similar to Bessel; since 1866 it has been seen as a craterlet surrounded by a light area. Pictures from Orbiter space-probes shows its form well, and it is not now believed that any real change occurred there.

Other features visible at this phase include the Lacus Somniorum (Lake of the Sleepers), which adjoins the Mare Serenitatis, near a large crater, Posidonius; the western part of the Mare Frigoris (Sea of Cold), together with two grand craters, Eudoxus and Aristoteles, 40 and 60 miles across respectively; and the smaller but very brilliant crater Menelaus, on the border of the Mare Serenitatis. The roughness of the southern part of the Moon is very obvious. Theophilus and its companions remain striking, but Langrenus and Petavius are less so, since they have lost most of their floor-shadow.

7th Day after New Moon. The Moon now appears as a half; this phase is known astronomically as *First Quarter*. Features which have come into view include some of the peaks of the Apennine range, the great plains Hipparchus and Albategnius, and many more formations in the southern uplands, such as the 90-mile, dark-floored Stöfler.

The Apennines are not the highest mountains on the Moon, since they rise to less than 20,000 feet, but they are certainly the most impressive.

151

Their peaks cast long shadows across the plain, and they are recognizable at a glance. The Apennine range forms part of the border of the Moon's most conspicuous sea, the Mare Imbrium (Sea of Showers), while further north may be seen part of the Mare Frigoris.

Half way along the terminator lie three great craters—Ptolemæus, Alphonsus and Arzachel—which form a chain. Ptolemæus, the northern member, is almost 100 miles across, and is extremely conspicuous when its floor is covered with shadow. Alphonsus, the crater inside which Kozyrev recorded a temorary reddish patch in 1958, adjoins Ptolemæus to the south, and is of comparable size. South again is Arzachel, smaller and deeper than its companions, and with a prominent central peak. Some way from the group may be seen another chain of three large formations—Walter, Regiomontanus and Purbach.

By First Quarter, the small but bright ray-system associated with Proclus, near the Mare Crisium, has become striking, though the more extensive systems of Tycho and Copernicus are not yet in view.

8th Day after New Moon. The Moon is now technically gibbous, and the terminator cuts through the well-defined plain of the Mare Imbrium. The Apennines and the adjacent range of the Alps are excellently placed. This is one of the most fascinating areas of the Moon; look for the Alpine Valley, which seems almost like a chisel-cut through the range, and for the three craters Archimedes, Aristillus and Autolycus and the Mare Imbrium. Archimedes, 50 miles in diameter, has a relatively level floor. On the terminator may be seen an even more interesting crater—Plato. It is 60 miles across, and is always recognizable because of the dark grey hue of its floor. At this phase, of course, it is filled with shadow.

The Ptolemæus chain is now some way from the terminator, but is unmistakable. Look too for the so-called Straight Wall, an 80-mile-long fault in the lunar surface near the small but prominent crater Birt. Much further south lies Clavius, one of the largest craters on the Moon; it is almost 150 miles in diameter, and contains other formations inside it. Clavius is situated in the rough southern uplands, but stands out because of its tremendous size. Binoculars will show it excellently.

9th Day after New Moon. Much of the Mare Imbrium is now to be seen, together with areas of the vast plain to the south—Mare Nubium, the Sea of Clouds. Near the boundary between the two the magnificent crater of Copernicus stands out, its floor covered with black shadow. Copernicus, 56 miles in diameter, has terraced walls and a complex central mountain mass, but is remarkable chiefly because of its system of bright rays. However, the rays are not well seen until nearer full moon.

Plato is striking; so too is Bullialdus in the Mare Nubium, which has been termed a smaller version of Copernicus (though it is not a ray-centre). The Straight Wall is visible as a dark line, and Clavius remains prominent. Not far from Clavius lies the major ray-centre of the Moon, Tycho; the crater itself is 54 miles across, and its bright walls make it easy to find even before its rays have come into view. South of the Clavius-Tycho region the surface is crowded with craters of all sorts and sizes.

10th Day after New Moon. The Mare Imbrium and Mare Nubrium are now wholly illuminated, and part of the Moon's largest 'sea'—Oceanus Procellarum, the Ocean of Storms—has come into view. One of the most superb features is the Sinus Iridum (Bay of Rainbows), bordering Mare Imbrium to the north-east. When the terminator passes closely to the west it juts out into the blackness, giving an appearance which has been termed the 'jewelled handle'. By now the Tycho rays are starting to become very noticeable, and we can also see another ray-crater, Kepler, not far from Copernicus.

At this stage in the lunation it is worth looking at craters such as Langrenus, Theophilus, Ptolemæus and Plato which have been described earlier. Plato's dark floor makes it easy to find, but the inexperienced observer will find it difficult to locate Langrenus and its companions at all now that they contain no appreciable floor-shadow.

11th Day after New Moon. The most interesting of the new features is Aristarchus, on the Oceanus Procellarum. It is less than 30 miles across, and only 5,000 feet deep, but it is the brightest object on the whole Moon; indeed, its brilliance has led to its being mistaken for an active volcano. Close beside it is its companion Herodotus, of similar size but not nearly so bright. Extending from Herodotus is a U-shaped winding valley. Temporary red patches have been seen in this area.

The ray-systems of Copernicus, Kepler and Tycho are striking, and it is worth noting the well-defined little Mare Humorum (Sea of Moisture). On the terminator, in the south-east, lies a vast crater known as Schickard, 134 miles in diameter.

Its walls are irregular in height, but one peak in them rises to almost 9,000 feet above the floor.

12th Day after New Moon. The shadows are becoming slight now; only along the eastern side of the Moon are they noticeable, and the Tycho and Copernicus rays are very prominent. Of the features newly brought into view, one of the most interesting is Grimaldi, a tremendous crater with rather low and irregular walls. It is not quite so large as Schickard—its diameter is 120 miles—but it is easier to locate, since its floor is the darkest portion of the Moon. It is worth comparing the hue of Grimaldi with those of Plato and Endymion. Close to Grimaldi is another large formation, Riccioli, which contains one very dark grey patch.

Full Moon. When the whole of the Moon's illuminated part is turned toward us, the shadows almost vanish, and much of the detail is drowned by the Tycho rays. It is fascinating to track these rays across the Moon; some of them extend for many hundreds of miles. The Copernicus rays are not so brilliant, but are prominent enough, as are the rays associated with Kepler and other centres. The outlines of the seas are perfectly clear, together with exceptional craters such as Aristarchus, Proclus, Grimaldi and Plato, but it is easy to see why full moon is an unsuitable time for the beginner to start trying to identify various formations.

After full moon, the phase decreases once more, and the shadows lengthen; the Mare Crisium, the Theophilus group, the Mare Serenitatis, Plato, the Mare Imbrium and all the other features are in their turn overtaken by lunar night, and the cycle begins once more with the following new moon.

A 3-inch telescope will show surprisingly fine lunar detail, and even a home-made spectacle-lens instrument will reveal many craters in addition to those which have been described here. Excellent photographic atlases have been produced, showing very delicate and complex detail. Even so it was only very recently that we had our first direct information about the 'other side of the Moon'. Until 1959, only 59 per cent of the surface had been available for study.

The Moon takes $27\frac{1}{3}$ days to revolve round the Earth. It spins once on its axis in precisely the same time, with the result that it keeps the same hemisphere turned permanently in our direction. Part of the Moon is always turned away from us, and can never be seen from the Earth. Effects known as *librations* result in a slight tilting to and fro, but 41 per cent of the surface is always averted.

Almost all the lunar craters are circular, though many of them have been ruined and distorted by later outbreaks. (Note, for instance, how Cyrillus has been deformed by the younger Theophilus.) Formations near the apparent centre of the disk, such as Ptolemæus, are seen 'full on'; those nearer the limb, such as Plato, are drawn out into ellipses by foreshortening. A good example of this effect is provided by the Mare Crisium. It appears elongated in a north-south direction, but in reality the east-west diameter is greater by 70 miles. Craters very near the limb appear as very long, narrow ellipses, and can never be properly studied.

The first manned landing on the Moon, achieved by the astronauts of Apollo 11 on 21 July 1969, was perhaps the greatest scientific triumph of all time, considered from a purely technical point of view. The landing was made in the Mare Tranquillitatis, not far from the twin craters Sabine and Ritter. Neil Armstrong was first to step out on to the lunar rocks, followed shortly afterwards by Edward Aldrin; Michael Collins remained in orbit round the Moon, though his rôle in the expedition was as important as those of his colleagues. The astronauts ventured out on to the surface, wearing protective spacesuits, and deposited scientific equipment, including a seismometer or 'moonquake-recorder'. A second expedition —Apollo 12, of November 1969—was equally successful; the astronauts were Charles Conrad, Alan Bean and Richard Gordon. Then, in 1970, came the near-disaster of Apollo 13, still fresh in our minds. Since then there have been four more Apollo missions, all of them highly successful. On the last three, the astronauts even took with them 'Moon cars' or Lunar Rovers, which were used to drive across the surface and thus extend the range of exploration considerably.

Much has been learned. The lunar rocks are of volcanic type, though in some ways they differ from the average terrestrial rocks; the age of the Moon is about the same as that of the Earth; there is no sign of life, either past or present. The surface is firm enough to bear the weight of a spacecraft, and there are no layers of deep, treacherous dust. Mild surface tremors do indeed take place, and records of them came back from the seismometers until September 1977.

The Apollo programme ended in December 1972, and it is not yet certain when manned exploration will be resumed; but meantime the Russians are carrying out investigations with their automatic 'Lunokhod' vehicles, which can crawl

157

around the surface, guided from Earth, and can send back information. If all goes well, we may expect a full-scale Lunar Base within the next few decades.

In spite of its unfriendliness, the Moon has a fascination all its own. It is our faithful companion, and it is without doubt the most specular object in the heavens from the viewpoint of the observer equipped with a pair of binoculars or a small telescope.

RECOGNIZING THE PLANETS

The planets, once known as 'wandering stars', are members of the Solar System. They revolve round the Sun at various distances and in various periods; some are larger than the Earth, others smaller; some are conspicuous, others faint. There are nine altogether, including our own world. Of these, Mercury and Venus are nearer to the Sun than we are, while the rest are more remote. Their sidereal periods, or revolution times, range from 88 days for Mercury up to nearly 248 years in the case of far-away Pluto.

The table on page 159 may be of general help.

It will be seen that the Solar System is divided into two groups of bodies. The four inner planets—Mercury, Venus, the Earth and Mars—are comparatively small and close to the Sun. Then comes a wide gap, now known to be occupied by numerous tiny worlds known as the *minor planets* or *asteroids*, followed by four giants—Jupiter, Saturn, Uranus and Neptune—and finally yet

PLANETARY DATA

Planet	*Mean distance from Sun, in millions of miles*	*Sidereal Period*	*Synodic Period*[1]	*Axial Rotation*	*Diameter in miles (equatorial)*	*Maximum Magnitude*
Mercury	36	88 days	115 days	58¼ days	3,000	−1·9
Venus	67	224·7 ,,	584 ,,	243 days	7,700	−4·4
Earth	93	365 ,,	—	23 h. 56m.	7,926	—
Mars	141·5	687 ,,	780 ,,	24 h. 37m.	4,200	−2·8
Jupiter	483	11¾ years	399 ,,	9¾h.	88,700	−2·5
Saturn	886	29½ ,,	378 ,,	10¼h.	75,100	−0·4
Uranus	1,783	84 ,,	370 ,,	±16h.	32,200	+5·6
Neptune	2,793	164¾ ,,	367½ ,,	±18h.	30,800	+7·7
Pluto	3,666	247¾ ,,	366¾ ,,	6d. 9h.	1,800	+14

[1] Interval between successive oppositions for superior planets, or conjunctions for Mercury and Venus.

another small planet, Pluto, which some astronomers tend to regard as being in a class of its own.

Mercury and Venus, known commonly as the *inferior planets*, have their own way of behaving and show phases similar to those of the Moon. When full, however, they are of course on the far side of the Sun, and will be unobservable with the naked eye; they will be above the horizon only during the hours of daylight.

Mercury never becomes conspicuous, since it is never very far from the Sun in the sky. Sometimes it may be seen in the west after sunset or in the east before dawn, shining like a bright star; but it is not easy to find, and the casual observer is not likely to notice it. In colour it is somewhat pinkish.

Venus is much larger than Mercury, and also much closer, so that it is much more prominent. It is actually the nearest body in the heavens apart from the Moon, and may come within 25,000,000 miles of us; at its brightest it may be seen in full daylight, and near elongation (maximum angular distance from the Sun) it is a striking object in the eastern or western sky. It is then so brilliant that it cannot be mistaken, and may even cast shadows. Oddly enough, it is at its best when at crescent phase. When favourably placed it may set as late as 5 hours after the Sun, or rise 5 hours before.

Venus is an 'evening star' when east of the Sun, and a 'morning star' when to the west. Dates of greatest elongation for the period 1983–85 are as follows:

Eastern elongations: 16 June 1983, 22 January 1985.

Western elongations: 4 November 1983, 13 June 1985.

Mars, which moves in an orbit beyond that of the Earth, does not show lunar-type phases, though it may appear gibbous. When it is opposite to the Sun in the sky it is said to be at *opposition*; it is then due south at midnight, and is well placed for observation. One year later the Earth has completed a journey round the Sun and has come back to its original position; but Mars moves more slowly, and has further to go. Before another opposition, therefore, the Earth has to catch Mars up. On an average it takes 780 days to do so, a period which is termed the *synodic periods* of Mars; there are, of course, slight fluctuations from the 780-day mean. It is approximately correct to say that oppositions of Mars occur every alternate year. Thus there was an opposition on 22 January 1978, and another on 25 February 1980, but there was no opposition in 1979, or in 1981.

Mars never approaches us much within 35,000,000 miles, and is a relatively small world, so that it is brilliant only for a few weeks to either side of opposition. It may then become extremely striking, both because of its brightness and because of the strong red colour which has earned it its name of 'the Planet of War'.

When some way from opposition Mars does look very much like a moderately conspicuous red star, and is easy to confuse with a real star. The only safe way to identify it is to check its position and then pick it out among the constellations. Watch it from night to night and you will see that it is shifting against the starry background. Its movement is naturally slow, but careful checking for several consecutive nights will reveal it. Like the other planets Mars always keeps within the Zodiac, which is a great help in identification.

Mars has an orbit which is less nearly circular

161

than that of the Earth; its distance from the Sun varies between 128,500,000 miles at its closest (*perihelion*) to 154,500,000 miles at its most remote (*aphelion*). When opposition occurs near perihelion Mars is at its best, since the distance is reduced. This was the case in 1956 and 1971, when for a few weeks Mars outshone every object in the night sky apart from the Moon and Venus. On the other hand the oppositions of 1980 and 1982 were unfavourable, since they took place when Mars was near aphelion. In those years the magnitude did not rise above -1. At its faintest, Mars may fall to the 2nd magnitude.

Jupiter is so large that it is conspicuous even across a distance of several hundred million miles. It is slightly yellowish and its brilliance makes it unmistakable; it can hardly be confused with any other object except perhaps Venus. Since it is slow-moving oppositions take place at intervals of only 13 months or so. Jupiter remains prominent for several months each year.

Saturn is less easy to recognize, and is often confused with a star. It is dull yellow in hue, and can never become much brighter than Capella or Arcturus. Its movements are even slower than those of Jupiter, and it comes to opposition each year.

Saturn was the outermost of the planets known in ancient times, and it was a surprise to astronomers when in 1781 a then unknown amateur named Herschel—afterwards Sir William Herschel—discovered the world we now know as Uranus. Despite its giant status Uranus is only just visible with the naked eye, and it is not easy to identify. A small telescope, however, shows that it is not in the least like a star; instead of appearing as a point of light it appears as a distinct greenish disk. At the present time it lies in Libra.

Two more remote planets have since been found: Neptune in 1846 and Pluto in 1930. We need not consider either in any detail, since telescopes are needed to show them. Neptune reveals a small bluish disk, but even the world's largest telescopes will not show Pluto as anything but a star-like speck.

An observer who has even a fair knowledge of the constellations will have no trouble in identifying Mars, Jupiter and Saturn, while Venus is always easy because of its brilliancy and because of its position relative to the Sun. Once a telescope becomes available it is clear that each of our neighbour worlds has its own particular points of interest, and planetary observation is one of the main concerns of the serious amateur astronomer.

THE PLANETS THEMSELVES

It cannot be said that binoculars or very small telescopes are of much use for making physical observations of the planets. The phases of Venus may be seen, and also the four chief moons or *satellites* of Jupiter, but that is practically all.

A 3-inch refractor, however, is enough to show many interesting features, including the polar caps of Mars and the superb ring system surrounding Saturn. In this book the best method will be to treat the planets one by one, and point out what may be observed with such an instrument.

Mercury merits little discussion. It is very elusive, and city-dwellers are unlikely to glimpse it at all. Our 3-inch refractor will show its phases, but certainly no markings on its small, pinkish disk. This is hardly surprising; Mercury is not a great deal larger than the Moon, and never comes much within 50,000,000 miles of us.

Without doubt Mercury is the most unwelcoming of the planets. It is almost without atmosphere, and it spins on its axis in just over 58 days. The surface temperature ranges between + 800° Fahrenheit at midday down to − 200° Fahrenheit during the long Mercurian night.

All serious observation of the planet from Earth must be done in broad daylight, since when visible to the naked eye Mercury is hopelessly low down; and this involves using a large telescope (preferably a refractor) fitted with accurate setting circles. Pictures sent back from the Mariner 10 probe in 1974 showed that Mercury is covered with craters very similar to those of the Moon. However, these craters cannot be seen from Earth, even with large telescopes; and to the observer it cannot be said that Mercury is a rewarding object.

Venus is very different. It is about the same size as the Earth, and is the most beautiful object in the heavens to the naked eye. Shining in the late dusk or early dawn, it resembles a floating lamp, and when visible near Christmas time it never fails to produce a large number of inquiries from people who believe it to be a return of the biblical 'Star of Bethlehem'.[1]

Unfortunately, Venus proves rather a dis-

[1] The Star of Bethlehem was not Venus or any other planet, and neither was it an ordinary star. No definite scientific explanation is available; it can hardly have been a nova or supernova, but there is a very faint chance that it was a bright comet. My own theory is that a bright meteor (or possibly two meteors) was the source of the legend.

appointment from the telescopic point of view. The phases are easy to see, but there is practically no surface detail; all that can be made out is the brilliant disk, devoid of any definite features. The reason for this is that Venus is surrounded by a dense atmosphere, and we can never see the planet's true surface. All we can observe is the upper part of the atmospheric layer. The various diffuse dusky shadings visible from time to time may be termed 'clouds', though they are quite unlike the clouds in the Earth's own atmosphere.

To make matters worse, Venus is not visible at all when it is at its closest to us. It is then at *inferior conjunction* and is approximately between the Earth and the Sun, so that its dark hemisphere is turned in our direction, and the planet is 'new'.

As the phase increases, so Venus draws further away from the Earth, and the apparent diameter shrinks. Moreover, the disk is so brilliant that it has to be studied against a bright background, as otherwise the elusive shadings are lost in the general glare from the disk. The best observations of Venus are made in full daylight, with the Sun above the horizon, but this is hardly an easy matter unless the telescope used is equipped with setting circles to locate the planet.

Incidentally, there are occasions when Venus passes directly between the Earth and the Sun; it is then seen as a black circle crossing the solar disk. Such *transits* are infrequent and the next will not occur until the year 2004. Transits of Mercury will take place on 12 November 1986 and 14 November 1999, but cannot be observed with the naked eye, since Mercury is smaller and more remote than Venus.

The atmosphere of Venus is by no means inviting. The clouds contain sulphuric acid, but the bulk of the atmosphere is made up of the gas known as carbon dioxide, together with various minor constituents. Carbon dioxide has a 'greenhouse' effect, and tends to trap the Sun's heat, so that the temperature of Venus' surface is high.

Various rocket probes have been sent to Venus—beginning with Mariner 2, of late 1962. The results have been a disappointment from the astronomical point of view. The surface temperature is about 900° Fahrenheit; the atmosphere is made up almost entirely of carbon dioxide, and the atmospheric pressure is 100 times as great as the pressure of the Earth's air at sea-level. In its way Venus is as hostile as the Moon, and we must discount any hope of finding life there. The length of the Venus 'day' is another surprise. It proves to be 243 Earth-days, which is longer than a Venus 'year'. This result has also been confirmed by radar measurements made from Earth. The first close-range pictures of the cloud-tops of Venus were sent back from Mariner 10 in 1974, and in 1975 and 1982 the Russians managed to obtain pictures of a rock-strewn surface, sent back by soft-landing probes.

Why is Venus so different from our own world? We have to admit that we do not know. It remains very much of a puzzle. Meanwhile, the owner of a small telescope will find it interesting to follow the phases, and to glimpse some of the fugitive cloudy shadings.

Mars is smaller than Venus, and is further away, so that it is well seen only for a month or two every alternate year. At such times it is a glorious naked-eye object, and a small telescope will show the principal markings on its disk; when it is more remote even large instruments will not reveal much.

Mars was always regarded as the one planet,

apart from Earth, upon which life might exist. Even in the earlier part of our own century there was support for the idea that the so-called 'canals', straight, artificial-looking lines crossing the Martian surface, were true waterways, constructed by intelligent beings. Yet even before the start of the Space Age this attractive idea had been discounted, and the most that could be expected was low-type plant life, covering wide areas and producing the celebrated dark markings. It was also known that as well as being thin, the Martian atmosphere was very deficient in water-vapour.

Direct exploration began in 1965, when the American probe Mariner 4 by-passed the planet and sent back pictures showing the surface to be cratered. The atmosphere was even more rarefied than had been thought; the main constituent was carbon dioxide, and it was assumed that the white polar caps were made up of solid carbon dioxide, though it has since been found that the basic caps are composed of ordinary water ice.

Further confirmation was obtained from Mariners 6 and 7, in 1969. Then, in 1971–2, Mariner 9 studied Mars from a closed orbit round the planet, sending back thousands of high-quality pictures which put an entirely new complexion upon matters. As well as being cratered, Mars showed deep valleys and giant volcanoes—one of which, Olympus Mons, rises to 15 miles above the mean surface level, and is crowned by a caldera 40 miles in diameter. There were also many features which looked suspiciously like dry riverbeds.

One major surprise was that the dark areas— such as the V-shaped Syrtis Major, familiar to all observers of the planet—are not depressed basins; some of them are high plateaux, and the idea of their being vegetation-tracts was finally given up.

167

All earlier ideas were thrown overboard. For instance there was the case of Hellas, a whitish, circular feature to the south of the Syrtis Major, which can at times become so brilliant that it looks almost like an extra polar cap. It had been assumed to be a lofty, snow-covered plateau; instead, it is the deepest basin known on Mars!

Then, in 1976, came the first successful soft-landings, with America's Vikings 1 and 2. (Earlier Russian attempts had been fruitless.) Viking 1 came down in the area known as Chryse, and at once began transmitting information. The area was rock-strewn; there were obvious sand-dunes, and the sky was salmon-pink rather than the expected blue, due to fine dust suspended in the atmosphere. The atmospheric pressure was a mere 7 millibars, and the winds ranged up to around 14 m.p.h. Viking 2 followed a few weeks later; this time the landing-site was in Utopia, further from the equator, and it was found that conditions there were similar, though the temperatures were lower. Even at noon in these parts of Mars, a thermometer still registers below $-20°$ F.

One of the most important aspects of the Viking programme was the search for life. Surface material was scooped up and drawn back for analysis in the hope of finding micro-organisms. Nothing positive was found, and yet there are plenty of problems concerning Mars which will not easily be solved. The 'riverbeds' seem to make up a definite drainage system associated with the huge volcanoes, and it is hard to resist the conclusion that they were cut by water; but no liquid water can exist there now, because the atmospheric pressure is too low. When the channels were cut, there must have been much more atmosphere than there is today. However, the lack of marked erosion means that on the geological time-scale the channels can hardly be very ancient; tens of thousands of years, but not millions. What, then, has happened to the atmosphere?

It has been suggested that Mars goes through 'fertile' periods, when all the volatiles in the polar caps are released into the atmosphere. This could be because of the changing tilt of the axis; if a cap is pointed sunward at the time of perihelion it would naturally receive more heat, and be more easily vaporized. Alternatively, there might be periods of intense vulcanism, when gases are sent out from below the crust to thicken the atmosphere temporarily. These and many other problems remain to be explained. At least we may be sure that even if there is no advanced life, Mars retains its unique interest.

There are two Martian satellites, both discovered by Asaph Hall in 1877. They have been named Phobos and Deimos. Both are very small, and are irregular in shape; Phobos, the larger, is shaped rather like a potato, and its greatest diameter is a mere 15 miles, while Deimos is even tinier. They were photographed from Mariner 9 and the Vikings, and have been found to be crater-pitted. Obviously they are faint objects as seen from Earth, and their nature is quite different from that of our own massive Moon.

Beyond Mars we meet with the minor planets, the junior members of the Sun's family. Ceres, the largest, has a diameter of only 700 miles; only one, Vesta, is ever visible without a telescope. It is thought that at least 40,000 minor planets exist, and recent Russian estimates increase this to nearer 200,000. Yet most of these worldlets are mere pieces of material a mile or two across, and are of no interest to the amateur. Their origin is

169

uncertain. Some astronomers believe them to be the remnants of a former planet which met disaster, though it is much more likely that they are cosmical débris which never collected together to form a large body.

A few of the minor planets have exceptional orbits which bring them well within the path of Mars. Eros, for instance, may approach the Earth to within a distance of 15,000,000 miles, and in 1937 a tiny body named Hermes passed by us at less than 500,000 miles. Icarus, named after the mythological flyer who went so close to the Sun that the wax fastening the wings to his body melted, has a perihelion distance of only 17,000,000 miles from the Sun, so that it must then be red-hot; Hidalgo travels from the orbit of Venus out almost as far as Saturn. Unfortunately all these bodies are so small that they are generally difficult to see even with large instruments.

Jupiter, giant of the Solar System, is big enough to contain over 1,300 globes the size of the Earth, and is a brilliant object in the night sky for several months each year. The owner of a moderate telescope will find it fascinating, and even field-glasses will show the four principal satellites.

Jupiter is not a hard, rocky body. Its surface is made up of rather dense, intensely cold gas, mainly hydrogen and hydrogen compounds such as ammonia and methane. Its internal constitution is not certainly known. Most authorities believe it to have a rocky core, surrounded by layers of liquid hydrogen which are in turn overlaid by the massive atmosphere. The internal temperature is certainly high. We have also found that there is a very strong magnetic field.

The disk is appreciably flattened at the poles, since the equatorial diameter (88,700 miles) considerably exceeds the polar diameter (82,800 miles). There is no mystery about this. Jupiter spins very rapidly on its axis, and centrifugal force results in a bulging out of the equatorial zone.

In a small telescope it is possible to see several of the bands which cross Jupiter, and are termed *belts*. Normally the darkest and broadest belt is the 'north equatorial'; the 'south equatorial' and the 'south temperate' belts may also be prominent, and a 3-inch refractor will often show a couple more. It has been found that the belts are caused by ammonia droplets which float in the upper part of Jupiter's gas-layer.

The belts are not perfectly straight and regular throughout their length, and under good conditions very intricate detail may be seen. Spots, too, are common, though generally short-lived. Now and then remarkable phenomena are observed; in 1959, for instance, the whole equatorial region showed a marked yellow–orange colour, unlike anything which had previously been recorded. The cause of this appearance remains unknown, but it gave the impression of being a high-altitude 'veiling' above Jupiter. By 1960 all was normal once more.

Jupiter rotates so quickly that a few minutes' observation will show a definite drift of the markings across the disk.

One important point is that the different features have different rotation periods; the equatorial zone has a period several minutes shorter than that valid for higher latitudes. This is extra proof, if proof indeed were needed, that the surface of Jupiter is not hard and solid.

No description of Jupiter would be complete without reference to the Great Red Spot, which seems to be in a class of its own, and is now known to be a kind of whirling storm. It became very prominent in 1878, and measured 30,000 miles long by 7,000 wide, so that its surface area was equal to that of the Earth; it was then of a strong red colour. Earlier drawings allow us to trace its history back as far as the year 1631. Since 1878 the Spot has faded, and at times (as in 1976) has become invisible, but it always comes back, and is frequently of a decided pinkish hue. It was well shown on the pictures sent back from the Pioneers of 1973 and 1974, and the Voyagers of 1979.

Of the 16 satellites, 12 are minute (the 14th has yet to be fully confirmed), but the remaining four are so bright that good binoculars will show them. A few people with exceptionally keen sight can see them without any optical aid whatsoever when conditions are favourable and Jupiter is high in the sky. Io and Europa are about the size of our Moon; Ganymede and Callisto are much larger, and have diameters of around 3,000 miles. All four 'Galilean' satellites were studied from the Voyager probes of 1979, and proved to be remarkable worlds. Io is the strangest of them; its surface is red —and showed active volcanoes, which was not in the least what had been expected. Ganymede and Callisto are heavily cratered.

The satellites revolve round Jupiter in periods which range from 1 day $18\frac{1}{2}$ hours for Io to 16 days $16\frac{1}{2}$ hours for Callisto. Since they revolve in the plane of the planet's equator they are usually seen more or less in a straight line. Their shifts from night to night are interesting to follow.

When a satellite passes in transit across Jupiter it may be seen against the planet's disk; if it passes behind Jupiter it is occulted for a period. The satellites may also enter Jupiter's shadow and suffer eclipse. It has been said that the four moons take part in 'celestial hide-and-seek', which is a graphic if unscientific way of putting it! Yearly astronomical almanacs give predictions of the various phenomena.

The second of the giant planets, and the outermost member of the Solar System known in ancient times, is Saturn. Since it is smaller than Jupiter, and much more remote, it is less conspicuous, but telescopically it is the gem of the sky. Its globe is very similar to that of Jupiter, and belts are seen as well as occasional spots; but its real glory lies in its system of rings.

Very small telescopes will indicate that there is something unusual about the appearance of Saturn. Higher powers show that this is due to what the Dutch astronomer Christiaan Huygens described 300 years ago as 'a flat ring, which nowhere touches the body of the planet, and is inclined to the ecliptic'. To be more accurate there are three main rings, two of which are bright, and the third (the Crêpe Ring) dimmer and semi-transparent.

The rings are not solid or liquid sheets, as used to be thought. A solid ring could not exist so close to Saturn; it would be disrupted by the powerful pull of gravity. It is now known that the rings are composed of numerous small particles moving independently round the planet in the manner of dwarf moons.

The two bright rings are separated by a gap known as Cassini's Division, in honour of the Italian astronomer who first observed it. This Division, 1,700 miles wide, appears as a dark line; it is due mainly to the gravitational influence of Saturn's innermost satellites. Under favourable conditions a 3-inch refractor will show it, but a second division (Encke's) in the outer ring is much more elusive.

The Crêpe Ring is comparatively obscure, and cannot be seen except with larger instruments.

The ring system measures 170,000 miles from side to side, but is only about 10 miles thick. Consequently the aspect of the planet changes from year to year, according to the angle from which the rings are seen. In 1974 and 1975, for instance, the rings were 'wide open' as viewed from Earth, and the spectacle was indeed glorious. By 1980 the rings were edge-on to us, and appeared as a thin line of light, vanishing completely except with powerful telescopes.

A scale model will show how thin the rings really are. If we represent Saturn by a 5-inch globe, the ring-span will be one foot, but the thickness will be only 1/5,000 of an inch—one-third the thickness of an ordinary sheet of writing-paper.

In every way Saturn is an intriguing object. The shadows of the rings on to the globe are not difficult to see and it is often possible to make out similar shadows cast by the globe on to the rings. Now and then unusual features appear on the disk itself; a white spot in the equatorial zone was discovered by W. T. Hay in 1933, and persisted for some weeks, while a similar though much less prominent white spot was observed in 1960. Moreover Saturn has a wealth of satellites. One of them (Titan) has a diameter of over 2,600 miles, and a dense atmosphere made up principally of nitrogen. A 2-inch refractor will show Titan well. Of the remaining attendants, a 3-inch will reveal Iapetus and Rhea, and a 4-inch is enough to show Dione and Tethys. The other moons—Mimas, Enceladus, Hyperion, Phœbe and the rest—are fainter.

Beyond Saturn we find two more giant planets.

Uranus, discovered by Herschel in 1781, also has a gaseous surface. It is faintly visible to the naked eye, and careful observation with a small telescope from night to night will reveal its motion against the background of stars. Large telescopes are needed to show any surface features, and all its five satellites are rather difficult objects. It is surrounded by a system of rings, but these cannot be seen with an ordinary telescope. They were discovered in 1977, when they passed in front of a star and made the star 'blink'.

Neptune was discovered in 1846, as a result of calculations made independently by U. J. J. Le Verrier in France and J. C. Adams in England. It is to all intents and purposes a twin of Uranus; it is slightly denser and has a slightly greater mass, but the two are strikingly similar. However, since Neptune is so remote it is naturally faint, and small telescopes will not readily show it as a disk. Of its two satellites, one (Triton) is over 3,000 miles across, and is brighter than any of the attendants of Uranus; the other (Nereid) is excessively faint.

Finally, on the edge of the planetary system, we find another small world—Pluto, tracked down by Percival Lowell of Martian canal fame, and first identified in 1930 by Clyde Tombaugh, working at the observatory which Lowell had founded. Since its discovery Pluto has proved consistently troublesome to the theorists. It is certainly smaller than the Earth, and is now believed to be smaller than the Moon, in which case it could not exert a measurable pull upon either Uranus or Neptune. Its orbit is relatively eccentric, and at the present time it is closer to the Sun than Neptune; not until 1999 will it again become 'the outermost planet'. In 1978 it was found to have a comparatively large satellite, now named Charon, so that Pluto may even be described as a double planet. An 8-inch reflector will show it, but it looks merely like a faint star.

There may well be another planet beyond Pluto, but it is bound to be very faint and difficult to locate. Yet the brilliant planets closer to us—Venus, Mars, Jupiter and Saturn—provide the amateur astronomer with a full observational programme, and enable him to carry out work which is of real use to his professional colleagues.

COMETS

A great comet is one of the most spectacular of celestial objects. It may be bright enough to be visible in full daylight, with a tail which stretches half way across the sky, and it will remain visible for some time—a few days or weeks—before fading into the distance. It is not surprising that ancient people believed comets to be portents of evil; in fact such fears are not quite dead even in our own century.

Yet a comet is not so important as it may look. It is made up of relatively small solid particles surrounded by an 'envelope' of very tenuous gas, while the tail also is made up of gas and what may be termed cosmic dust. Clearly, then, a comet is quite unlike a planet. Even if the Earth happened to pass straight through a comet's head no permanent damage would result; on two occasions during the last hundred years or so our world has actually passed through the tail of a comet without coming to the slightest harm.

Unfortunately, great comets have been rare of late. Two were visible in 1910, and there have been none since. We can guarantee that a fairly bright comet will be seen in 1986, since this object—Halley's Comet—comes back every 76 years; but we do not know whether we shall meet with another before then. Bennett's Comet of 1970 was one of the brightest of recent years, though it could not compare with the brilliant comets of the nineteenth century. West's Comet of 1976 was also bright.

It seems certain that all comets are members of the Solar System and revolve round the Sun, but in most cases their orbits are different from those of the planets. Instead of being nearly circular a cometary orbit is markedly elliptical. For instance Encke's Comet, which has a period of 3·3 years, travels from inside the orbit of Mercury well out into the asteroid zone, while Halley's Comet penetrates far beyond remote Neptune. We do know of a few comets which have nearly circular paths, but all these are extremely faint.

A comet shines only because of the influence of the Sun, and so cannot be seen unless it is reasonably close to the Sun and to the Earth. For instance, Halley's Comet, due at perihelion in February 1986, was not recovered until 1982; it was then well over 1,000 million miles away and so faint that only the world's largest telescope could record it.

Halley's Comet is a regular visitor, appearing every 76 years. Accounts of it go back well before the time of Christ. It was the comet visible in 1066, and is said to have struck terror into the hearts of Harold's Saxons; it was back in 1835 and 1910. No other bright comet has a period of less than several centuries, and the numerous *short period comets*, which take only a few years to complete one journey round the Sun, are too faint to be seen without optical aid.

It is also important to note that the conventional picture of a comet, with shining head and impressive tail, is true only for the 'great' comets. Those of short period often appear as fuzzy blurs not unlike luminous cotton wool, with no tails whatsoever. They are easy to mistake for star clusters or nebulæ; in fact Charles Messier drew up his now famous catalogue of nebulæ so that he could have a quick reference list of 'objects to avoid' during his searches for new comets.

According to a theory now much in favour, a comet's head is made up chiefly of particles composed of 'ices'. When the comet nears the Sun these ices begin to evaporate and material is released; it is this material which forms the tail. It is a well-known fact that a comet will develop a conspicuous tail only as it approaches perihelion, if indeed it develops a tail at all.

It is not true to say that a comet always travels head-first. The particles in the tail are of such a size that they are particularly affected by what is called 'solar wind', streams of particles coming from the Sun. The particles are of atomic size but are electrified, and tend to 'push away' the tenuous matter in a comet's tail. The result of this action is that the tail always points outward; when a comet has passed perihelion, and has begun its return journey, it moves tail-first.

Apart from Halley's, all brilliant comets move in orbits so eccentric that the periods amount to many centuries—or even, in some cases, to thousands or millions of years. To all intents and purposes we may regard these comets as non-periodical. Once they have made their visit to the

Sun they draw back once more into the depths of space, and will not be seen again for many generations. We cannot predict when they will appear; they are unexpected visitors, and take us by surprise.

Many faint comets, too, move in eccentric paths, and may be classed as non-periodical. This applied, for example, to the two naked-eye comets of 1957, Arend-Roland and Mrkos. The first of these, discovered by Arend and Roland in the early part of the year, became quite conspicuous, though it was very minor compared with a 'great' comet; it was easily visible without telescope, and attracted a good deal of attention. In particular it was unusual in having a curious kind of 'spike', extending from the head in a direction opposite to that of the tail, due to material spread out in the comet's orbit being illuminated by the Sun and seen at a suitable angle. The comet discovered in the autumn of 1957 by the Czech astronomer Antonin Mrkos, and independently found shortly afterwards by a 15-year-old British amateur named Clive Hare, was about equally bright, but remained a naked-eye object for only a few days.

Yet another naked-eye comet, Burnham's, was seen in April 1960, in the region of Ursa Major and the north celestial pole. Unfortunately it proved to be rather disappointing, and was none too easy to see without optical aid, though binoculars gave a clear view of it. The Seki-Lines Comet of May 1962, expected to become brilliant, also failed to come up to expectations.

Another interesting modern comet was that of autumn 1965, also discovered by the Japanese amateurs Ikeya and Seki. As seen from some parts of the world (notably Arizona), it made a brilliant

179

showing, but its glory was brief, and from Europe it never became visible to the naked eye, so that it too must be classed as something of a disappointment in view of its exceptional promise. Then came Bennett's Comet of 1970, which was bright enough to be really spectacular—though British observers did not see it to best advantage, since when at its most brilliant it was in the far south of the sky. Kohoutek's Comet, expected to become really brilliant in late 1973, proved to be a great disappointment, though it did become visible with the naked eye. Its revolution period is of the order of 70,000 years. West's Comet of 1976 was more spectacular; as it moved outward after passing perihelion it showed signs of disintegration. It also has an extremely long period.

Powerful telescopes will usually show at least a couple of comets in the sky at any particular moment, but comets brilliant enough to be of interest to the casual sky-watcher have been rare in recent times. Let us hope that it will not be too long before another spectacular wanderer moves in toward the Sun, making a brave show for a while before returning to the depth of space from whence it came.

SHOOTING-STARS

Finally, in our survey of the natural bodies of the Solar System, we come to the shooting-stars or *meteors*, some of which shine brilliantly for a second or two before destroying themselves in the Earth's upper air.

Most celestial objects, including comets, seem to move very slowly in the heavens. A comet appears almost stationary with respect to the star-patterns; it shares in the diurnal movement around the celestial pole, but it has to be watched over a period of hours for its individual movement to become detectable. Should an observer see a body flash across the sky at express speed, he will know that what he is seeing is not a comet but a meteor.

A meteor is a small particle revolving round the Sun in the manner of a dwarf planet. Normally it is too faint to be seen, but if it comes close to the Earth it will enter the upper air, and will brush so violently through the air particles that friction will be set up. This friction produces heat, and the meteor is destroyed in a streak of radiation. It is this appearance which we term a shooting-star; and clearly there is no similarity between a real star, which is a sun, and a shooting star, which is generally smaller than a grape.

When a meteor encounters the Earth it may be moving at a relative speed of anything up to 45,000 m.p.h., and it will become luminous at a height of over 100 miles above ground level. Most meteors are destroyed by the time they have fallen to 60 or 70 miles, and in this case they finish their journey in the form of very fine 'dust'. It is worth noting that a grape-sized meteor ranks as a giant; the average meteor is much more insignificant still.

Larger bodies are encountered now and then, big enough to fall to the ground without being destroyed. Most science museums have collections of these *meteorites*, which range from pebble-sized objects to vast blocks. The heaviest known meteorite is still lying where it fell, at Hoba West in Africa, and has an estimated weight of 60 tons; another tremendous object fell in Arizona in pre-historic times, and caused a mile-wide crater which has become a popular tourist attraction. In 1908 a large meteorite fell in Siberia, blowing pine-trees flat for 20 miles in all directions from the impact point, and there was another major fall in 1947 in the Vladivostok area. The last known English meteorite fell at Barwell (Leicestershire) on Christmas Eve, 1965. It seems, however, that there is a fundamental difference between meteorites and the much smaller ordinary shooting-stars, and large meteorites are very rare.

When the Earth passes through a shoal of meteors, as happens regularly, the result is a shower of shooting-stars. Because the meteors are really travelling through space in parallel paths, the shooting-stars of any particular shower will seem to come from a definite point or *radiant*; this is a perspective effect, just as an observer will see that parallel roads appear to meet at a point toward his horizon. The shower is named according to the constellation in which the radiant lies; thus the August *Perseids* radiate from Perseus, the December *Geminids* from Gemini, and so on.

Of the various annual showers, the following are the most important:

The Quadrantids come from the Boötes–Ursa Major region, and are so named because the old star-maps contain a now rejected constellation there, Quadrans Muralis (the Mural Quadrant). The positions of the other radiants are obvious from the names.

The Leonids used to produce magnificent displays, those of 1833 and 1866 being spectacular; it is said that 'meteors fell as thick as snowflakes'. A comparable display occurred in 1966, but was seen only from the Western Hemisphere, as it took place during daylight over Europe. The richest reliable shower nowadays is that of August, radiating from Perseus. Anyone who stares up

IMPORTANT ANNUAL METEOR
SHOWERS

Name	Normal Limits: beginning	end	Remarks
Quadrantids	Jan. 3	Jan. 4	Usually a sharp maximum, January 4
Lyrids	Apr. 19	Apr. 22	Moderate shower; swift meteors
Aquarids	May 1	May 13	Very swift meteors with long paths
Perseids	July 27	Aug. 17	The richest annual shower
Orionids	Oct. 15	Oct. 25	Moderate shower; swift meteors
Taurids	Oct. 26	Nov. 16	Not usually a rich shower
Leonids	Nov. 15	Nov. 17	Unpredictable
Geminids	Dec. 9	Dec. 13	Good, rich shower

at the sky for several consecutive minutes during a night in early August will be unlucky not to see at least one shooting-star, though in some years the presence of the Moon drowns all but the brighter meteors.

A really brilliant meteor, known—most misleadingly—as a fireball, may produce a trail which persists for some minutes; even ordinary shooting-stars may leave brief trails.

For ordinary meteor-watching no optical instruments are needed. The path of the object may be tracked among the stars, and its magnitude estimated on the stellar scale, together with any special features such as fluctuations in brightness. Though the showers appear regularly, they do not account for all the meteors observed. Many shooting-stars are *sporadic*, and are not members of swarms, so that they may appear from any direction at any moment.

It used to be thought that meteors came from

interstellar space, but this theory has now been disproved. They are genuine members of the Solar System, and are associated with comets. This association was shown, in dramatic fashion, during the last century. In 1823 the Austrian amateur astronomer Biela discovered a comet which proved to revolve round the Sun in a period of $6\frac{3}{4}$ years. It presumably returned to perihelion in 1839, but was not observed because of its unfavourable position in the sky. When it came back once more in 1846, it surprised astronomers by dividing into two separation portions. The twins returned on schedule in 1852, were missed in 1859 because they were badly placed, and were expected once more in 1865-66. Yet they failed to put in an appearance; despite the most careful searches no trace of them could be found. In 1872, when Biela's Comet should have been back again, the comet itself was absent, but in its place appeared a shower of shooting-stars. For years afterwards meteors were seen each year at the time when the Earth crossed the path of the disintegrated comet. A few of these 'Bieliid' shooting-stars are still seen annually about 28th November, though the shower has now become very feeble.

We do not know for certain how meteors were formed. They may represent the débris left over when the planets were born, but we do not even know how the planets themselves came into being. It used to be supposed that they were pulled off the Sun by the tidal action of a passing star, but this theory is mathematically unsound, and it seems much more likely that the planets were formed from a 'solar nebulæ'—a cloud of dust and gas associated with the Sun.

At any rate a rich meteor shower is both interesting and spectacular. Even more so, perhaps, are the rarer 'fireballs', which may shine more brilliantly than the Moon during their dash across the heavens. Early August is certainly the best time for meteor-watching, but at any time during the year there is always the chance of seeing a display of celestial fireworks.

MAN-MADE MOONS

Until 1957 the Earth had only one satellite—the Moon. Of course there is still only one natural attendant, but in recent years the Moon has been joined by various other bodies made on Earth and launched into space by means of rocket power. All the *artificial satellites* are small, astronomically speaking, but since they are visible from Earth they must be included in any book dealing with what used to be called 'star-gazing'.

So long as we remain on the Earth's surface, we must be content to carry out our observations by looking through the whole thickness of the blanket of atmosphere. Unfortunately the air is dirty and unsteady; moreover, there are layers in the upper regions which block out some of the radiations as effectively as a sheet of metal will block the beam of a torch. In consequence astronomers are anxious to carry out observations from above the top of the atmosphere. Now that manned space-flight is well under way, the prospects are good. There is every reason to suppose that a full-scale orbiting astronomical observatory will have been launched soon.

Already we have collected a tremendous amount of information. For instance, there are certain objects in the sky which send out X-rays; these

X-rays cannot be studied from ground level, but instruments in artificial satellites have pin-pointed the sources with amazing accuracy. But for full information we must await either the orbital observatory, or else an observatory set up on the surface of the Moon. In many ways the latter may eventually prove to be the better proposition.

Machines such as balloons, propeller aircraft and jets can operate only when surrounded by atmosphere; above a comparatively few miles there is so little air that normal methods of flight are useless. However, rockets are not so limited. They function by *reaction*, and are at their best in vacuum, since air resistance is eliminated.

After the war, rocket vehicles powered by liquid propellants were used to send instruments into the upper atmosphere. Valuable results were obtained, but it became clear that the ordinary rocket is limited in scope. It could remain aloft for only a few minutes at best, and then fell back to the ground, destroying itself and often destroying its instruments as well—despite various refinements such as the breaking away and parachuting down of the instrument containers.

The American research workers therefore determined to launch an artificial satellite, but were forestalled by the Russians. On October 4 1957 Soviet scientists launched Sputnik I, a sphere weighing 184 lb., which entered an orbit round the Earth and at once began to transmit radio signals which were heard by operators all over the world. Since then many other satellites have been similarly launched, and probe vehicles have been sent to Venus, Mars, Mercury and Jupiter. More may be expected in the coming years.

Once an artificial satellite has been sent up, and put into a stable orbit round the Earth, it will not come down unless it moves within the atmosphere; it will behave as a natural astronomical body. If, however, it enters the upper air at any point in its orbit, it will be affected by friction and will eventually drop into the lower, denser air, so that it will be destroyed in the same fashion as a meteor. This was the fate of most of the early satellites, including Sputnik I. On the other hand modern 'communications satellites' are to all intents and purposes permanent, since they lie beyond the effective atmosphere.

An artificial satellite does not look in the least like a balloon hanging in the sky, as some people expected. It gives the impression of a starlike point moving steadily and obviously against the background of true stars. Up to 1982 the most spectacular satellites were the two American Echos, which were nothing more than large balloons coated with reflective material, and were launched to act as passive communications satellites.

Amateurs can do useful work in satellite tracking. No equipment is needed other than a reliable stop-watch, plus a good knowledge of the constellations. The method is to note the time when a satellite makes up a definite configuration with some known stars, so that its position may be plotted on a map. Of course it is essential to know the sky really well; it is not really useful to record a satellite as passing near, say, β Cancri when it is actually near the region of η Leonis!

There are satellites of many kinds—meteorological, geodetic and so on. And, needless to say, they have been of tremendous importance in communications; today it is commonplace for us to tune in to the BBC television service and have a direct view of something which is happening in, say, New York!

The first man in space, Yuri Gagarin of the U.S.S.R, made a full circuit of the Earth on 12 April 1961. Since then there have been flights by both American and Russian pioneers; at the time of writing the 'endurance record' is held by the main crew of a Salyut space-station, who remained aloft for over 6 months. Weightlessness is not an immediate danger, though there are uneasy doubts that prolonged periods of zero gravity may be harmful.

The Apollo moon-flights were in the nature of reconnaissance expeditions. It is true to say that the mission of Apollo 13, in the following year, came as an unpleasant shock; it served as a reminder that space-flight is a very hazardous business indeed. The last four Apollo missions were highly successful, but it is too early to say just when it will be possible to send a man to Mars. I suspect that it may have to await the development of nuclear-powered rockets, which should be practicable at some time during the 1980s.

Probes have already been sent to the inner planets and to Jupiter and Saturn. Voyager 2 is now on its way to Uranus and Neptune. If all goes well, we may hope for detailed information from these remote worlds within the next few years. Certainly events have moved much more quickly than most people anticipated only a few years ago. It is not so long since the very idea of reaching the Moon was regarded as science fiction; but science fiction has been transformed into science fact. The Space Age has well and truly begun, and the last part of the twentieth century bids fair to be eventful by any standards.

Astronomy is an ever-changing science, and one which transcends all national considerations; observers from all countries work together in the common quest for knowledge. Moreover, everyone can take an interest in it—even the casual star-gazer who has no profound scientific knowledge and has no ambition to undertake serious research work. Those who take the trouble to learn the constellations, recognize the planets, and perhaps obtain a small telescope to see the wonders of the heavens for themselves, will find that they have not wasted their time.

APPENDIX

ASTRONOMICAL SOCIETIES

The British Astronomical Association (secretarial address: Burlington House, Piccadilly, London W.1.) is the premier observational society in the United Kingdom; it has existed since 1890, and has an impressive record. No technical qualifications are necessary for membership. A regular journal is published, and there are various observing sections for those who want to undertake useful work. Monthly meetings are held at Savile Row in London.

In addition, many larger towns and cities (Manchester, Birmingham, Leicester, Liverpool and Norwich, for instance) have their own astronomical societies which meet regularly. These are listed in the annual *Yearbook of Astronomy*.

Membership of some organized body is strongly recommended. The beginner will not only learn but will come into contact with those who have similar interests, with results which can hardly fail to be beneficial.

BIBLIOGRAPHY

Many astronomical books have been published during recent years, and to list them all would require many pages. Therefore only a very small selection is given.

S. Mitton (1977) *Exploring the Galaxies*, Faber & Faber.

Patrick Moore (1976) *The Amateur Astronomer*, Lutterworth; (1977) *Guide to the Moon*, Lutterworth. *The Guiness Book of Astronomy Facts and Feats* (1979). *The Unfolding Universe* (Michael Joseph 1982).

Iain Nicolson (1976) *Simple Astronomy*, Nelson; (with Moore) (1974) *Black Holes in Space*, Orbach & Chambers.

INDEX

191